# Maths from Scratch
# for Biologists

# Maths from Scratch for Biologists

Alan J. Cann
*University of Leicester, UK*

JOHN WILEY & SONS, LTD
Chichester • New York • Weinheim • Brisbane • Singapore • Toronto

*Other Wiley Editorial Offices*

John Wiley & Sons Inc., 111 River Street, Hoboken, NJ 07030, USA

Jossey-Bass, 989 Market Street, San Francisco, CA 94103-1741, USA

Wiley-VCH Verlag GmbH, Boschstr. 12, D-69469 Weinheim, Germany

John Wiley & Sons Australia Ltd, 33 Park Road, Milton, Queensland 4064, Australia

John Wiley & Sons (Asia) Pte Ltd, 2 Clementi Loop #02-01, Jin Xing Distripark, Singapore 129809

John Wiley & Sons Canada Ltd, 22 Worcester Road, Etobicoke, Ontario, Canada M9W 1L1


*Library of Congress Cataloging-in-Publication Data*

Cann, Alan.
  Maths in biology/Alan J. Cann.
    p. cm.
  Includes bibliographical references (p.).
  ISBN 0-471-49834-3 (cased) — ISBN 0-471-49835-1 (pbk.)
    1. Biomathematics. ☐ I. Title.
  QH323.5.C363 2002
  570′.1′51 — dc21                          2002028068

*British Library Cataloguing in Publication Data*

A catalogue record for this book is available from the British Library

ISBN 0 471 49834 3 Hardback
     0 471 49835 1 Paperback


Typeset in 11/13.5pt Sabon by Thomson Press (India) Ltd., Chennai
Printed and bound in Great Britain by TJ International, Padstow, Cornwall
This book is printed on acid-free paper responsibly manufactured from sustainable forestry,
in which at least two trees are planted for each one used for paper production.

*To my Mum and Dad*

# Contents

# Preface

This book arose from my own need for a text that I would be happy to recommend to my students. Although there is no particular shortage of volumes claiming to help biologists with mathematics, all those I am familiar with have one of two flaws. Either they are written by well-meaning mathematicians and pay scant attention to biology, or they are not appropriate for the level at which most of the problems lie – new college students who do not have much confidence in approaching mathematical problems, in spite of extensive prior exposure to mathematics in school.

I make no claim to be a mathematical genius. Indeed, I believe my struggle to explain the material in an easily accessible form is one of the strengths of this book, bringing me closer to the students I am trying to communicate with. I reject any charges of 'dumbing down' – anyone who has ever tried to help a panic-stricken student in the grip of maths phobia will know that a calming but not patronizing voice is an essential attribute in these circumstances. Throughout, my intention is to provide a highly accessible text for students who, with or without formal mathematics qualifications, are frightened by the perceived 'difficulty' of mathematics and unwilling, inept or inexperienced in applying mathematical skills. To accommodate these students, many of whom opt to undertake studies in biology in the belief (conscious or unconscious) that this is a way of pursuing a scientific career while avoiding mathematics, the ethos of the book is consciously informal and intended to be confidence-building.

The maths in this volume has been checked vigorously, but I cannot guarantee that the text is entirely free of numerical errors. In addition there may be some passages where the subject matter is not expressed as clearly as I would have hoped. I rely on readers to point these out to me – as I am sure they will.

**Alan J. Cann**
*University of Leicester, UK.*
alan.cann@le.ac.uk

# 1

# Maths in Biology

*Mathematics, from the Greek, manthano, 'to learn'*

Some people opt to undertake studies in biology in the belief (conscious or unconscious) that this is a way of pursuing a scientific career while avoiding maths. This book is designed to be accessible to students who, with or without formal mathematical qualifications, are frightened by the perceived 'difficulty' of maths and hence are unwilling to apply what mathematical skills they might have. Have you ever noticed when you have been taught how to solve a mathematical problem, that you still don't know *why* you need to do a particular step? This is the root of many problems with maths, so this book will try explain the *why* of maths, in addition to the *how*. Sometimes, these explanations may seem unnecessary, but I urge you not to skip them – understanding why you need to do something is the key to remembering how to do it. The intention is to be informal and confidence-building to ensure that *all* readers will gain a general appreciation of basic mathematical, statistical and data handling techniques appropriate to biology. I will try to explain the jargon which confuses the non-numerically minded.

In subsequent chapters, we will look at manipulating numbers, units and conversions, molarities and dilutions, areas and volumes, exponents and logs and statistics. However, the basic advice in this chapter is really the most important part of the book, so please keep reading.

## 1.1. What can go wrong?

It is easy to make mistakes with maths. One answer looks much like another, so how can you tell if it is right or not? Look at some examples of the sort of mistakes it is all too easy to make. Everyone knows that numbers are meaningless without the units which define what they mean (more of this in Chapter 3). Even if we avoid the elementary mistake of forgetting this and giving an answer of '33.6' (33.6 what? volts? metres? frogs?), things are not always simple. Consider the following questions:

> An aquarium has internal dimensions of $100 * 45 * 45$ cm. What is its volume in litres?

This is fairly simple. Calculate the volume in cubic centimetres then convert to litres. 1 litre $= 1000 \, cm^3$ so divide by 1000:

$$100 * 45 * 45 = 202\,500 \, cm^3 \div 1000 = 202.5 \text{ litres}$$

However, life is not always that simple. If the same calculation is given in a different way, it is not as easy to answer:

> An aquarium has internal dimensions of $39 * 18 * 18$ inches. What is its volume in litres?

This is harder because the units in which the data is given and those in which the answer is required are from different systems of measurement. In real life, this happens all too frequently.

# WARNING!
## Using mixed units is dangerous (see Chapter 3).

To avoid mistakes we need to convert the units so that they are consistent throughout. However, this means there are two ways to do the calculation:

1. Convert inches to centimetres (1 inch $= 2.54$ cm), then perform the calculation as above,

$$(39 * 2.54) * (18 * 2.54) * (18 * 2.54)$$
$$= 99.06 * 45.72 * 45.72$$
$$= 207\,066.94 \, cm^3 \div 1000 = 207.067 \text{ litres}$$

2. Calculate the volume in cubic inches, then convert to litres (1 cubic inch = 0.0164 litres, so conversion factor = 0.0164):

$$39 * 18 * 18 = 12\,636 * 0.0164 = 207.23 \text{ litres}$$

In general, the best method is the one which requires fewer conversions and fewer steps (b). However, this depends on what conversion factors you have to hand – if you have to calculate a conversion factor from cubic inches to litres, it may be better to use (a). Note that the accuracy of conversions from one unit to another depends on the number of significant figures used. Significant figures are: 'the minimum number of digits needed to write a given value (in scientific notation) without loss of accuracy'. The **most** significant figure is the left-most digit, the digit which is known most precisely. The **least** significant figure is the right-most digit, the digit which is known least precisely.

Significant figures are important when reporting scientific data because they give the reader an idea of how accurately data has been measured. Here are the rules:

1. All **non-zero numbers** (1, 2, 3, 4, 5, 6, 7, 8, 9) are always significant, e.g. 12 345 has five significant figures; 1.2345 also has five significant figures.

2. All **zeroes between non-zero numbers** are always significant, e.g. 10 002 has five significant figures; 1.0002 also has five significant figures.

3. All **zeroes which are to the right of the decimal point** *and* **at the end of the number** are always significant. This rule sometimes confuses people since they cannot understand why. The reason is because these zeros determine the accuracy to which the number has been calculated, e.g. 1.2001 has five significant figures; 12 000 has two significant figures; 1.0200 has five significant figures (here the 'placeholder' zero to the right of the decimal point is significant because it is between non-zero numbers).

4. All other zeroes are not significant numbers, e.g. 1 000 000 has one significant figure (the zeros are just 'placeholders'); 1 000 000.00 has three significant figures (the 1 and the two zeros at the end of the number); 0.0200 has three significant figures (the 'placeholder' zero to the right of the decimal point is not significant since it is not between non-zero numbers); 1 000 000.01 has nine significant figures (zeros between non-zero numbers).

Using the appropriate number of significant figures in calculations is important, since it prevents loss of accuracy. However, computers and calculators frequently give ridiculously large numbers of significant figures – way beyond the accuracy with which a measurement could be made. For this reason, and for ease of performing calculations (particularly when estimating, see below), it is often necessary to 'round off' the number of significant figures in a number. Note that this is 'rounding off', *not* 'rounding up', which leads to inaccuracy and errors. 'Rounding up' a digit which is followed by a 5 (e.g. 5.45 becomes 5.5) introduces errors in calculations because the digits one, two, three and four are 'rounded down' (four possibilities) but the digits five, six, seven, eight and nine are all 'rounded up' (five possibilities). 'Rounding off' avoids this error:

1.  If the digit following the figure that is to be the last digit is **less than 5**, drop it and all the figures to the right of it.

2.  If the digit following the figure that is to be the last digit is **more than 5**, increase by 1 the digit to be rounded, i.e. the preceding figure.

3.  If the digit following the figure that is to be the **last digit is 5**, round the preceding figure so that it is even.

## Examples

Round 123.456789 to three significant figures = 123 (rule a: round the number off)

Round 123.456789 to five significant figures = 123.46 (rule b: round the last digit up)

Round 123.456789 to four significant figures = 123.4 (rule c: make the last digit even)

Round 123.356789 to four significant figures = 123.4 (rule c: make the last digit even)

Round 123.456799 to eight significant figures = 123.45680 (note that 9 rounds up to 10, not down to 0).

## 1.2. Estimating

> **Whenever you have calculated an answer, always make a rough estimate to see if your answer is sensible and to avoid mistakes.**

Calculators and computers spit out numbers at the press of a key, but are the answers right? Estimating is a vital skill if you wish to become confident and proficient with numbers. However, estimating and calculating are **not** the same thing and it is important to understand the difference. Where calculation attempts to produce the most accurate answer possible (within limits of experimental error), estimation deliberately avoids accuracy in order to simplify working out the answer.

1. If the question is $6 * 5$ and your calculated answer is 4, could this possibly be correct? Could the answer be *less* than the numbers multiplied together?

2. If you are asked to solve an equation for $x$ (Chapter 2) and your answer is $7x$, something is wrong.

3. When you calculate the answer to $6.42213 \div 2.36199$ to six significant figures (2.71895), make an estimate to one or two significant figures to check: $6 \div 2 = 3$, so 2.71895 looks right, whereas 27.1895 looks wrong.

If you have used a computer or calculator to calculate an answer, it is best to work out the estimate in your head or on a scrap of paper in order to check for any errors you may have introduced by using the machine. This is why estimation involves simplifying the calculation – an estimate is not meant to be accurate, but it should be easy to calculate and a reliable check. Aside from performing the calculation, *estimating* is the most important part of ensuring that answers to problems are correct. Some calculations in biology are complex and involve many steps (Chapter 5). Estimating is particularly important here to ensure the answer looks sensible. Manipulation of numbers and equations may not give a numerical answer but a mathematical term (e.g. $3y - 2$). Here, the trick is to check your answer by substituting back into the original equation to see if it works (Chapter 2).

## 1.3. How to use this book

If you have been told to use this book as part of a particular course, you had better follow the instructions given by whoever is running the course. Other than that, you can use this book however you want. Some people may want to read though all (or most) of the chapters in order. Others may skip sections and dip into chapters that they feel they need. Either way is fine, as long as you can solve problems consistently and accurately and, most importantly, that you gain the knowledge and confidence to start to try to work out possible answers.

## 1.4. Mathematical conventions used in this book

To make them easier to read, numbers with more than four digits are split into groups of three digits separated by spaces (not commas), e.g. 9 999 999 is nine million, nine hundred and ninety nine thousand, nine hundred and ninety nine. I have also chosen to use the asterisk (*) as a multiplication sign rather than '×' or a dot, since these are sometimes confusing when written.

# 2

# Manipulating Numbers

*Algebra (from the Arabic, al-jabr, 'the reduction') – a form of maths where symbols are used to represent numbers*

**LEARNING OBJECTIVES:**

On completing this chapter, you should be able to:

- understand the basic rules of algebra;
- perform simple algebraic manipulations;
- identify and manipulate fractions.

Arithmetic is concerned with the effect of operations (e.g. addition, multiplication, etc.) on specified numbers. In algebra, operations are applied to variables rather than specific numbers. Why? Here is a classic example:

John is 10 years old. His father is 35 years old. After how many years will the father be twice as old as the son?

You could try to find the answer by experimenting with different numbers, but this is laborious. The better way is to treat this an algebra problem and write the problem as an equation which we can then solve.

Let the father be twice as old as the son in $x$ years time. The son will then be $(10 + x)$ years old and the father will be $(35 + x)$ year old:

$$2(10 + x) = 35 + x$$

Therefore,

$$20 + 2x = 35 + x$$

Simplify this by subtracting $x$ from each side to keep the equation balanced:

$$20 + x = 35$$

Simplify by subtracting 20 from each side to keep the equation balanced:

$$x = 15 \text{ years (son is 25 and father is 50)}$$

## 2.1. Manipulating numbers

To manipulate numbers, you need to know the rules. In mathematics, this is known as the 'Order of Operations' – an internationally agreed set of arbitrary rules which allows mathematicians the world over to arrive at the same answers to problems:

**BEDMAS**

Order of operations – the order in which operations are performed:

Brackets (work from the inside out)
Exponents (see Chapter 6)
Division
Multiplication
Addition (left to right)
Subtraction (left to right)

In algebra, there are two sorts of statement which you need to be able to recognize:

1. A mathematical expression is a string of symbols which describes ('expresses') a (potential) calculation using operators (symbols indicating an operation to be performed, e.g. plus, minus, divide, etc.) and operands (symbols which the operators act on), e.g.

$$2x + y$$

Expressions do not contain an equal sign, but can often be simplified, that is converted to a simpler form containing fewer terms.

2. A mathematical equation contains an equal sign. The terms (groups of numbers or symbols) on both sides of the equal sign are equivalent, e.g.

$$2x = y$$

You can do anything you want to an equation, as long as you treat both sides equally. To solve an equation, you must find the values(s) of the variable(s) which make the equation true, that is both terms equal. A mathematical formula also represents a relationship between two or more variables (symbols or terms whose values may vary) and/or constants (numbers or terms whose value is fixed), e.g.

$$e = mc^2$$

A formula is simply an equation which expresses a rule or principle as symbols, i.e. the recipe which allow you to calculate the value of the terms.

## 2.2. Solving equations

To 'solve' an equation, you must find the value(s) of the variable(s) which make the equation 'true', i.e. makes the terms on either side of the equal sign equal. There are seven steps to follow in order to solve an equation ('**BICORS**'):

1. **Brackets** – if an equation contains brackets ('**B**'; also known as parentheses, which group symbols together), solve these first. Multiply each item inside the bracket by the symbol just outside the bracket.

2. **Isolate** – move all the terms containing a variable to the same side of the equal sign ('isolate' the variable; '**I**').

3. **Combine** – combine like terms, that is if an equation contains more than one term containing the same variable (e.g. $z$), combine them ('**C**').

4. **Opposite** – for each operator in an equation, perform the opposite process ('**O**'), for example, if the equation contains a minus sign, add, or if it contains a multiplication sign, divide.

5. **Reduce** – reduce ('**R**') fractions to their lowest terms (e.g. $33/11 = 3/1 = 3$).

6. **Substitute** – finally, *always* check your answer by putting this value back into the original equation ('substitute' for the variable; '**S**').

You will not always have to perform all of these steps, depending on the equation. For example, if an equation does not contain any brackets, just move on to the next step, but do go through all the steps in order. Solving equations often involves simplifying the expressions they contain, which means getting all similar terms (e.g. $x$) on the same side of the equal sign. All of this sounds more complicated than it actually is and is best illustrated by some examples.

## Examples

Solve for $x$ (i.e. find the value of $x$ that makes the equation true):

$$4x = 2(6x) - 4$$

Expand *Brackets* (**B**icors):

$$4x = 12x - 4$$

Simplify to *Isolate* (b**I**cors) the variable:

$$4x - 12x = -4$$

*Combine* (bi**C**ors) like terms:

$$-8x = -4$$

Carry out the *Opposite* (bic**O**rs) process:

$$8x = 4$$

Divide by the coefficient of the variable (variable $= x$, coefficient $= 8$):

$$8x/8 = 4/8$$

Simplify the equation by *Reducing* (bico**R**s) fractions to their lowest terms:

$$x = 1/2$$

Check the answer by *Substituting* (bicorS) it back into the original equation:

$$4(1/2) = 2[6(1/2)] - 4$$
$$2 = 2(3) - 4$$
$$2 = 6 - 4$$
$$2 = 2$$

Solve for $x$ (find the value of $x$):

|  |  |
|---|---|
|  | $5(x-4) = 20$ |
| **B** (*5): | $5x - 20 = 20$ |
| (**I, C**)O ( + 20): | $5x = 40$ |
| O(R) (/5): | $x = 40/5 = 8$ |
| S: | $5(8-4) = 20$ |
|  | $5(4) = 20$ |
|  | $20 = 20$ |

Solve for $z$ (find the value of $z$):

|  |  |
|---|---|
|  | $z/4 + 4 = 16$ |
| (**B, I, C**)O ( − 4): | $z/4 = 12$ |
| O(R) (*4): | $z = 48$ |
| S: | $48/4 + 4 = 16$ |
|  | $12 + 4 = 16$ |
|  | $16 = 16$ |

Note that equations do not always have a numerical answer – sometimes the value of the variable can only be expressed in terms of its relationship with another variable:

Solve for $x$ (find the true value of $x$):

|  |  |
|---|---|
|  | $2x + 4 = 2y + 4$ |
| (**B, I, C**)O ( − 4): | $2x = 2y + 4 - 4$ |
| **R** (/2): | $2x = 2y$ |
|  | $x = y$ |
| S: | $2y + 4 = 2y + 4$ |
|  | $2y = 2y$ |
|  | $y = y$ |

Solve for $t$ (find the true value of $t$):

$$t + 5 = x$$
$$(\mathbf{B}, \mathbf{I}, \mathbf{C})\mathbf{O} \; (-5): \quad t = x - 5$$
$$(\mathbf{R})\mathbf{S}: \qquad\qquad x - 5 + 5 = x$$
$$x = x$$

There are two main sorts of equation:

1. Linear equations – equations where the exponents of all the variables (powers of the variable, see Chapter 6) are equal to 1 and there is no multiplication between variables. Graphs of linear equations plot as straight lines, e.g. $y = 2x + 3$.

2. Non-linear equations – equations where the exponent (power, see Chapter 6) of one or more of the variables is not equal to 1 or there is any multiplication between variables. Graphs of non-linear equations plot as curves. This includes all polynomial functions [e.g. $f(x) = 4x^3 + 3x^2 + 2x + 1$], such as:

   • quadratic equations, e.g. $x^2 + 5x + 6 = 0$;
   • cubic equations, e.g. $x^3 + bx^2 + cx + d = 0$, etc.

Although non-linear equations are common in biology, this chapter is primarily concerned with linear equations. Many people find the idea of solving equations difficult. The answer to this is to practise. For this, you can use the problems at the end of this chapter. When you become more confident, you can move on to non-linear equations. Word problems are particularly useful to help you think through what you are being asked, but can be surprisingly difficult for some people. In real life, information is frequently presented in this form rather than as an equation. The trick is to start by converting words into numbers. Again, this is a skill that you can acquire by practice – use the problems at the end of this chapter.

## 2.3. Why do you need to know all this?

You need to know all this because you cannot go very far in biology without encountering topics like enzyme kinetics.

## Example

In an enzyme-catalysed reaction, the reactant (S) combines reversibly with a catalyst (E) to form a complex (ES) with forward and reverse rate constants of $k_1$ and $k_{-1}$, respectively. The complex then dissociates into product (P) with a reaction rate constant of $k_2$ and the catalyst is regenerated:

$$E + S \underset{k_{-1}}{\overset{k_1}{\rightleftharpoons}} ES \overset{k_2}{\longrightarrow} E + P$$

From this can be derived the Michaelis–Menten equation:

$$v = \frac{V_{max} * [S]}{K_m + [S]}$$

where $v$ = reaction rate (velocity), $[S]$ = substrate concentration, $V_{max}$ = maximum rate, $K_m$ = Michaelis–Menten constant = substrate concentration at half the maximal velocity ($V$), i.e. $K_m = [S]$ when $V = V_{max}/2$.

$K_m$ measures enzyme/substrate affinity – a low $K_m$ indicates a strong enzyme – substrate affinity and vice versa. However, $K_m$ is not just a binding constant that measures the strength of binding between the enzyme and substrate. Its value includes the affinity of substrate for enzyme, but also the rate at which the substrate bound to the enzyme is converted to product (see Table 2.1).

For ribonuclease, if $[S] = K_m = 7.9 * 10^{-3}$ M, then substituting into the Michaelis–Menten equation:

$$v = \frac{V_{max} * 7.9 * 10^{-3}}{7.9 * 10^{-3} + 7.9 * 10^{-3}}$$

Table 2.1  $K_m$ for various enzyme reactions

| Enzyme | Reaction catalysed | $K_m$ (M) |
|---|---|---|
| Chymotrypsin | Ac-Phe-Ala + $(H_2O)$ → Ac-Phe + Ala | $1.5 * 10^{-2}$ |
| Carbonic anhydrase | $HCO_3^- + H^+$ → $(H_2O) + CO_2$ | $2.6 * 10^{-2}$ |
| Ribonuclease | Cytidine 2′,3′ cyclic phosphate + $(H_2O)$ → cytidine 3′-phosphate | $7.9 * 10^{-3}$ |
| Pepsin | Phe-Gly + $(H_2O)$ → Phe + Gly | $3 * 10^{-4}$ |
| Tyrosyl-tRNA synthetase | Tyrosine + tRNA → tyrosyl-tRNA | $9 * 10^{-4}$ |
| Fumarase | Fumarate + $(H_2O)$ → malate | $5 * 10^{-6}$ |

Simplifying this by dividing the top and bottom of this equation by $7.9 * 10^{-3}$,

$$K_m = [S], \quad V = V_{max}/2$$

so

$$v = \frac{V_{max} * 1}{2}$$

When $[S] = K_m$, $v = V_{max}/2$ and hence the Michaelis–Menten equation works.


## 2.4. Fractions

When you perform algebraic manipulations, you soon encounter fractions, which means parts of numbers. We all learned to manipulate fractions in school, but in these days of computers and calculators, many people have forgotten how to do this. Remembering how to multiply and divide fractions causes particular problems.

All fractions have three components – a numerator, a denominator and a division symbol:

$$\frac{Numerator}{Denominator}$$

The division symbol in a simple fraction indicates that the entire expression above the division symbol is the numerator and must be treated as if it were one number, and the entire expression below the division symbol is the denominator and must be treated as if it were one number. The same order of operations (BEDMAS) applies to fractions as to other mathematical terms. Brackets instruct you to simplify the expression within the bracket before doing anything else. The division symbol in a fraction has the same role as a bracket. It instructs you to treat the quantity above (the numerator) as if it were enclosed in a bracket, and to treat the quantity below (the denominator) as if it were enclosed in another bracket:

$$\frac{(Numerator)}{(Denominator)}$$

In a simple fraction, the numerator and the denominator are both integers (whole numbers), e.g.

$$\frac{1}{2}$$

A complex fraction is a fraction where the numerator, denominator or both contain a fraction, e.g.

$$\frac{1/2}{3}$$

To manipulate (e.g. add, subtract, divide or multiply) complex fractions, you must first convert them to simple fractions.

A compound fraction, also called a mixed number, contains integers and fractions, e.g.

$$\frac{4 - 1/2}{3}$$

As with complex fractions, to manipulate compound fractions, you must first convert them to simple fractions.

No fraction (simple, complex or compound) can have a denominator with an overall value of zero. This is because, if the denominator of a fraction is zero, the overall value of the fraction is not defined, since you cannot divide by zero. A numerator is allowed to take on the value of zero in a fraction, although any legitimate fraction (denominator not equal to zero) with a numerator equal to zero has an overall value of zero. If there is a single minus sign in a simple fraction, the overall value of the fraction will be negative. If there is an even number of minus signs in a simple fraction, the value of the fraction is positive. If there is an odd number of minus signs in a simple fraction, the value of the fraction is negative, e.g.

$$\frac{-5}{-6} = 0.83 \quad \text{and} \quad \frac{-5}{6} = -0.83 \quad \text{but} \quad \frac{-3-6}{4} = \frac{-9}{4} = -2.25$$

## 2.5. The number 1

Although it may seem obvious, the number 1 has several properties which can easily be overlooked. This becomes important when working with fractions:

1. Multiplying any number by 1 does not change the value of the number. Dividing any number by 1 does not change the value of the number.

2. The number 1 can take many forms, e.g. $4-3=1$ or $10-9=1$. These can be substituted for 1 because they have the same value.

3. When the numerator of a fraction is equivalent to the denominator of a fraction, the value of the fraction is 1. This only works for legitimate fractions, i.e. where the denominator does not equal zero. You can substitute such a fraction with the number 1.

4. You can express any integer as a fraction by dividing by 1, or by choosing a numerator and denominator so that the overall value is equal to the integer.

5. To factor an integer, break the integer down into a group of numbers whose product equals the original number. Factors are separated by multiplication signs. Note that the number 1 is the factor of every number. All factors of a number can be divided exactly into that number.

## 2.6. Lowest common multiple and greatest common factor

When manipulating fractions you frequently need to find these two terms – you will see examples of this below. To find the lowest common multiple (LCM) of two numbers, make a table of multiples (e.g. 2 and 3; see Table 2.2). To find the greatest common factor (GCF), meaning numbers or expressions by which a larger number can be divided exactly ('factoring'), make a table of factors (e.g. 8 and 12; see Table 2.3).

Table 2.2  Table of multiples

| Multiples of 2 | Common multiples | Multiples of 3 |
| --- | --- | --- |
| 2 | – | 3 |
| 4 | – | – |
| 6 | 6 | 6 |
| – | – | 9 |

**Table 2.3**  Table of factors

| Factors of 8 | Common factors | Factors of 12 |
|---|---|---|
| 1 | 1 | 1 |
| 2 | 2 | 2 |
| – | – | 3 |
| 4 | 4 | 4 |
| 8 | – | 6 |
| – | – | 12 |

# 2.7. Adding and subtracting fractions

To be able to add or subtract fractions to or from fractions, the denominators must be the same ('common'): $\frac{1}{2} + \frac{3}{8}$ cannot be added, but $\frac{4}{8} + \frac{3}{8}$ can.

To find a common denominator so you can add or subtract fractions, find the LCM of all the denominators involved. Then, make the denominators equal the LCM by multiplying both the denominator and numerator by the corresponding factor of the LCM. Whenever you manipulate fractions, the final step is to reduce the answer to the lowest terms:

1. Factor the numerator.

2. Factor the denominator.

3. Find the fraction mix that equals 1.

*Example*

Add

$\frac{4}{3} + \frac{2}{5}$   The LCM of 3 and 5 (the denominators) is 15.

$\frac{4}{3*5} + \frac{2}{5*3}$   Both denominators must equal the LCM, so multiply 3 by 5, and 5 by 3. Now both denominators are the same ('common').

$\frac{4*5}{15} + \frac{2*3}{15}$   To avoid altering the problem, multiply the numerators by the same factor as their respective denominators. This is the same as multiplying each fraction by 15/15, i.e. by 1.

$\frac{20}{15} + \frac{6}{15}$   Now the denominators the same, add the fractions together.

$\frac{26}{15}$   You cannot reduce this fraction further, i.e. reduce numerator and denominator to their LCFs, so this is the final answer.

To build a fraction is the reverse of reducing the fraction. Instead of searching for one in a fraction and reducing, insert a 1 (or equivalent) and 'build' the fraction:

$$\frac{2}{2} = \frac{2}{3} * 1 = \frac{2}{3} * \frac{4}{4} = \frac{8}{12}$$

You might do this to find a common denominator so that you can add or subtract fractions.

## 2.8. Multiplying fractions

1. To multiply fractions, multiply the numerator(s) by the numerator(s) and the denominator(s) by the denominator(s).

2. *Do not cross-multiply* [i.e. do not multiply the numerator(s) by the denominator(s)].

3. After finding the product of the fractions, reduce the product to its simplest form.

4. You can multiply more than one fraction in a single step – simply multiply all the numerators, then all the denominators and then reduce the result to the lowest common terms.

5. To multiply a whole number and a fraction, convert the whole number to a fraction, multiply the numerators, multiply the denominators and reduce the results.

*Example*

$$\frac{3}{4} * \frac{6}{7} = \frac{18}{28} = \frac{18/2}{28/2} = \frac{9}{14}$$

## 2.9. Dividing fractions

'Cross-multiplication' is involved in dividing fractions (i.e. multiplying the numerator by the denominator). If you wish, you can do this by taking the reciprocal of the second fraction (i.e. flip the numerator and the denominator) and multiplying numerator with numerator and denominator with denominator.

You cannot divide more than one fraction in a single step. If multiple fractions are involved, break the problem down in separate parts.

To divide a fraction by a whole number, or a whole number by a fraction, first convert the whole number to a fraction.

*Example*

$$\frac{6}{2/3} = \frac{6/1}{2/3} = \frac{6}{1} * \frac{3}{2} = \frac{18}{2} = \frac{9}{1} = 9$$

## 2.10. Fractions, decimals and percentages

It is easy to convert a fraction to a decimal – just type the fraction into a calculator, divide the numerator by the denominator and read the result as a decimal. Example:

$$60/150(= 6/15 = 2/5) = 0.4$$

However, it is not always easy to use this method to calculate the exact value of fractions, for example try using your calculator to find the exact value of $22/7$. It is not possible to use a calculator to convert a decimal into a fraction, so this is done by dividing the fraction by 1, then multiplying the result by 1 in a form that will remove the decimal.

*Example*

$$3.3 = \frac{3.3}{1} = \frac{3.3}{1} * \frac{10}{10} = \frac{33}{10}$$

Most people are able to calculate percentages properly, but as mistakes are sometimes made, it is perhaps worth covering this briefly. 'Per cent' means 'for each 100', e.g. 25% means 25 out of 100. To calculate a percentage you need to know the number in the entire (total) group and the number in the part of the group you are interested in ('subgroup').

$$\% = \frac{\text{subgroup}}{\text{total}} * 100$$

**Table 2.4**  Common conversions from percentages to decimals and fractions

| Percentage | Decimal | Fraction | Reduced |
|---|---|---|---|
| 10% | 0.1 | 10/100 | 1/10 |
| 20% | 0.2 | 20/100 | 1/5 |
| 25% | 0.25 | 25/100 | 1/4 |
| 30% | 0.3 | 30/100 | 3/10 |
| 33.33% | 0.3333 | 33.33/100 | 1/3 |
| 40% | 0.4 | 40/100 | 2/5 |
| 50% | 0.5 | 50/100 | 1/2 |
| 60% | 0.6 | 60/100 | 3/5 |
| 66.66% | 0.6666 | 66.66/100 | 2/3 |
| 70% | 0.7 | 70/100 | 7/10 |
| 75% | 0.75 | 75/100 | 3/4 |
| 80% | 0.8 | 80/100 | 4/5 |
| 90% | 0.9 | 90/100 | 9/10 |
| 100% | 1.0 | 100/100 | 1 |

The mistake which is sometimes made is to divide the number in the subgroup by the remainder rather than by the total. To convert a percentage to a fraction, place the number over a denominator of 100, then reduce to lowest terms.

*Example*

$$45\% = \frac{45}{100} = \frac{9}{20}$$

The conversions in Table 2.4 are very common and should probably be memorized.

## 2.11. Ratios and proportions

Like percentages, ratios (pronounced 'ray-sho' or 'ray-she-oh', from the Latin for 'calculation') are often to used make comparisons between similar things, that is things measured in the same units. Ratios tell you how one item is related to another and can be written in several different ways, for example, in a population consisting of seven white mice and nine brown mice, the ratio of white to brown animals can be written as:

1. A fraction – 7/9.

2. In words, using 'to' – seven white mice **to** nine brown mice.

3. With a colon – 7:9.

As with fractions, for clarity ratios should always be reduced to their lowest terms. For example, the ratio 24:48 is equal to 12:24, etc., i.e. 1:2. A ratio can express the relationship between more than two items, for example, in a cake recipe containing three spoons of treacle, four spoons of sugar and nine spoons of flour, the ratio of treacle to sugar to flour is 3:4:9. Note that the numbers in this ratio do not tell you how big the cake is going to be, just the relative amounts of ingredients.

To tell if two ratios are equal, use a calculator and divide. If division gives the same answer (called the 'quotient' – the number obtained by dividing one number by another) for both ratios, then they are equal. A quotient is therefore a single number, whereas a ratio is a pair of numbers used to make a comparison. A proportion is a statement of equality between two ratios, e.g.

$$1/2 = 3/6$$

When one of the four numbers in a proportion is unknown, cross-multiplication may be used to find the unknown number ('solve the proportion').

### Example

Solve for $x$:

$$\frac{1}{2} = \frac{x}{4}$$

By cross-multiplying (as when dividing fractions, Section 2.9):

$$1 * 4 = 2 * x = 4$$

so

$$2 * x = 4$$

Dividing both sides by 2,

$$x = 4/2 = 2$$

A common difficulty with ratios is knowing which way round to write them. This depends on what question is being asked, for example, in our group of seven white mice and nine brown mice, the proportion of brown mice is 7:9 and the proportion of white mice is 9:7. The convention is to write the group which is being referred to first, so the molar ratio of hydrogen to oxygen in water ($H_2O$) is 2:1 and the molar ratio of oxygen to hydrogen in water is 1:2. Ratios crop up in everyday calculations all the time and are common in biology, but perhaps are most frequently encountered in calculations involving molarities and dilutions (Chapter 4). You can use the problems at the end of this chapter to practise calculations involving ratios.

## Problems (answers in Appendix 1)

### Basic equations

2.1. Solve for $x$ (find the value of $x$):     $3x + 3 = 5$

2.2. Solve for $y$:     $5y + 12 = 22$

2.3. Solve for $z$:     $10z + 9 = 6$

2.4. Solve for $w$:     $8w + 8 = 12$

2.5. Solve for $h$:     $9h + 86 = 99$

2.6. Solve for $a$:     $77a - 75 = 1$

2.7. Solve for $B$:     $11B + 11 = 11$

2.8. Solve for $\phi$:     $123\phi - 1353 = -123$

2.9. Solve for $f$:     $4f + 12 = 17$

2.10. Solve for $x$:     $2x = 3x - 2$

2.11. Solve for $x$:     $4x = 2x - 3$

2.12. Solve for $p$:     $3p = p + 6$

2.13. Solve for $z$:     $4z = 2z - 5$

2.14. Solve for $a$:     $22a = 41a - 38$

2.15. Solve for $x$:     $(x/6) = (x/2) + (5/4)$

2.16. Solve for $t$:     $(t/3) = (t/6) + (1/3)$

2.17. Solve for $w$:     $(2w/3) + 3 = (w/4)$

2.18. Solve for $m$:     $(100m/3) + (22/33) = (101m/3)$

## Multiple variable equations

**2.19.** Solve for $x$ (find the value of $x$):     $x + 9 = y$

**2.20.** Solve for $B$:     $10B + 2 = z - 6$

**2.21.** Solve for $n$:     $3n + 6 = x - 10$

**2.22.** Solve for $y$:     $2y + 3 = 2x + 3$

**2.23.** Solve for $y$:     $5y + 6 = 2x$

**2.24.** Solve for $c$:     $c - 8 = 4z + 2$

## Word problems

**2.25.** The sum of nine plus twice a number equals twenty-three. What is the value of the number?

**2.26.** Forty-four mealworms are placed in an escape-proof bowl in a lizard's cage. In twenty-four hours, the lizard visits the bowl to feed three times, eating the same number of mealworms each time. At the end of the experiment, eight mealworms remain uneaten. How many mealworms does the lizard eat on each visit to the food bowl? (Hint: the difference between forty-four and three times a number is equal to eight.)

**2.27.** The spine of a mammal contains forty-three vertebrae, the same number in each of the cervical, thoracic, lumbar, sacral and coccygeal regions, plus an additional three vertebrae. How many vertebrae are present in each region of the spine of this species?

**2.28.** A total of nine results are available from an experiment performed by students. Six students performed the experiment on Tuesday and each obtained a result. The rest of the class performed the experiment on Wednesday, but only one-third of these students obtained a result. How many students performed the experiment on Wednesday?

**2.29.** Nine birds of prey raised as part of a captive breeding programme are tagged with radio transmitters and released into the wild. One year later, only four birds are still alive. How many of the birds released have died?

**2.30.** Sheila starts the term with £240 in her bank account. Each week she withdraws the same amount of money. After six weeks, her account is £90 overdrawn. How much money did she withdraw each week?

**2.31.** A researcher counts eleven male robins with successful breeding territories in an area of $2400\,m^2$. Then they observe another successful breeding male in the same area. What is the average area occupied by each breeding male?

**2.32.** Nine students go out for a meal. In the restaurant each student puts the same amount of money on the table to pay for the meal. The bill is £142 and £2 is left over. How much money did each student put on the table?

**2.33.** At pH 7.4 an enzyme transforms 45 mmol of substrate $min^{-1}$. At the sub-optimal pH of 6.9, the enzyme transforms 12 mmol of substrate $min^{-1}$ less than at pH 7.4. How much substrate is transformed per minute at pH 6.9?

## Fractions

**2.34.** Calculate $\frac{3}{2} + \frac{2}{3}$

**2.35.** Calculate $\frac{7}{9} - \frac{5}{6}$

**2.36.** Calculate $\frac{8}{9} * \frac{5}{7}$

**2.37.** Calculate $\frac{6}{7} - \frac{5}{6}$

**2.38.** Express as a fraction $1 + \frac{6}{8}$

**2.39.** Express as a fraction $2 + \frac{1}{12}$

**2.40.** Express as a fraction $2 + \frac{8}{11}$

## Ratios

**2.41.** Are 4:12 and 36:72 equal ratios?

**2.42.** Seventy-five per cent of the prey items captured by bats are moths. Express the amount of moths in a bat's diet as a ratio.

**2.43.** If $x = 6$ and the ratio of $x:y = 2:5$, what is the value of $y$?

**2.44.** If a swallow flies 100 metres in 10 seconds, how long would it take it to fly 15 400 metres?

**2.45.** What volume of a $5\,g\,L^{-1}$ solution is required to make 100 mL of a $2\,g\,L^{-1}$ solution?

See Chapter 4 for more problems of this type.

# 3

# Units and Conversions

**LEARNING OBJECTIVES:**

On completing this chapter, you should be able to:

- describe the seven principal SI units;
- understand how other SI units are derived from the principal units;
- be able to apply SI units in calculations.

Measurement is fundamental to science. Observations or calculations without accompanying units of measurement are meaningless. In the course of history many different systems of measurement have arisen. While it is possible to convert between different measurements, this is inconvenient and results in inaccuracies, and frequently mistakes. Consequently, it was recognized that there was a need for a universal system of measurement.

## 3.1. The SI system of units

The metric system was invented by the French Academy of Science in 1799 as a 'rational' system of measurement – part of the 'Age of Enlightenment' which followed the French Revolution. Originally, it consisted of three units and was known as the 'MKS' system:

1. The **metre** – 1/10 000 000 the distance between the pole and the equator.

2. The **kilogram** – the mass of a standard platinum cylinder kept at the International Bureau of Weights and Measures in Paris.

3. The **second** – $1/(24 * 60 * 60)$ of a day.

This was gradually extended and rationalized and in 1960, the so-called SI system ('Système International d'Unités') was adopted internationally. Other systems of measurement persist to some extent, notably the 'foot, pound, second' (FPS) system, also known in the United States as the English Engineering System (EES), but the SI system is now the indisputable language of science.

A dimension is an abstract quality of measurement without scale (e.g. length, mass, time, etc.). A unit is a number which specifies a previously agreed scale (e.g. metres, kilograms, seconds). In the SI system, there are seven fundamental dimensions, measured by seven principal units, as shown in Table 3.1. Many other measurements are derived by combining the seven principal units, as shown in Table 3.2.

**Table 3.1**  Principal SI units

| Dimension | Unit | Abbreviation | Definition |
|---|---|---|---|
| Amount of a substance | mole | **mol** | The amount of substance that contains as many elementary units as there are atoms in 0.012 kg of carbon-12 |
| Electric current | ampere | **A** | The current which produces a specified force between two parallel wires 1 m apart in a vacuum |
| Length $(L)$ | metre | **m** | The distance light travels in a vacuum in 1/299 792 458 s |
| Luminous intensity | candela | **cd** | The intensity of a source of light of a specified frequency, which gives a specified amount of power in a given direction |
| Mass $(M)$ | kilogram | **kg** | The mass of an international prototype in the form of a platinum–iridium cylinder kept in Paris |
| Time $(T)$ | second | **s** | The time taken for 9 192 631 770 periods of vibration of the caesium-133 atom to occur |
| Temperature | kelvin | **K** (not °K) | 1/273.16 of the thermodynamic temperature of the triple point of water |

Table 3.2   Derived SI units

| Derived unit | SI unit | Abbreviation | Definition |
|---|---|---|---|
| Acceleration | – | – | $m\,s^{-2}$ |
| Area | – | – | $m^2$ |
| Density | – | – | $kg\,m^{-3}$ |
| Frequency | hertz | **Hz** | $cycles\,s^{-1}$ |
| Velocity | – | – | $m\,s^{-1}$ |
| Viscosity | – | – | $kg\,m^{-1}s^{-1}$ |
| Volume | – | – | $m^3$ |

The litre (L, the volume of 1 kilogram of pure water at 4°C and a pressure of 760 mm of mercury) is a metric unit but not an SI unit, since all SI units are derived from the principal units, hence the SI unit of volume is the $m^3$. Other commonly used abbreviations include h (hour, but not H, Hr or hr) and g (gram – do not confuse with *g*, the acceleration due to gravity).

## 3.2. SI prefixes

One of the most frequent criticisms of SI units is that many are either too large or too small for convenient use. In some cases this may be true. For this reason, SI (and metric) units can be made larger or smaller by the use of appropriate prefixes (Table 3.3). Some of these will already be familiar, such as the *kilo*metre and the *milli*litre, others less so. Other than the $10^1$, $10^2$, $10^3$ and $10^{-1}$, $10^{-2}$, $10^{-3}$ prefixes, which represent 10-fold changes, all the other SI prefixes involve 1000-fold increases or decreases in scale – an adjustment of three decimal places. This makes it much easier to remember the absolute magnitude of the largest and smallest SI prefixes, e.g. giga $= 10^9$, femto $= 10^{-15}$, etc.

The main advantage of the interlocking SI units is that they make it easy to convert from one scale to another, e.g. for a cube with sides of length *x*:

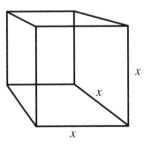

Table 3.3  SI prefixes

| Name | Symbol | Size | Factor |
|------|--------|------|--------|
| yotta | Y | 1 000 000 000 000 000 000 000 000 | $10^{24}$ |
| zetta | Z | 1 000 000 000 000 000 000 000 | $10^{21}$ |
| exa | E | 1 000 000 000 000 000 000 | $10^{18}$ |
| peta | P | 1 000 000 000 000 000 | $10^{15}$ |
| tera | T | 1 000 000 000 000 | $10^{12}$ |
| giga | G | 1 000 000 000 | $10^{9}$ |
| mega | M | 1 000 000 | $10^{6}$ |
| kilo | k | 1000 | $10^{3}$ |
| hecto | h | 100 | $10^{2}$ |
| deca | da | 10 | $10^{1}$ |
|  |  | 1 |  |
| deci | d | 0.1 | $10^{-1}$ |
| centi | c | 0.01 | $10^{-2}$ |
| milli | m | 0.001 | $10^{-3}$ |
| micro | μ | 0.000 001 | $10^{-6}$ |
| nano | n | 0.000 000 001 | $10^{-9}$ |
| pico | p | 0.000 000 000 001 | $10^{-12}$ |
| femto | f | 0.000 000 000 000 001 | $10^{-15}$ |
| atto | a | 0.000 000 000 000 000 001 | $10^{-18}$ |
| zepto | z | 0.000 000 000 000 000 000 001 | $10^{-21}$ |
| yocto | y | 0.000 000 000 000 000 000 000 001 | $10^{-24}$ |

If $x = 1\,m$, the volume of the cube $= 1*1*1 = 1\,m^3$. Since $1\,m = 100\,cm$, the volume of the cube in centimetres $= 100\,cm * 100\,cm * 100\,cm = 1*10^6\,cm^3$. Since $1\,m = 1*10^6\,\mu m$, the volume of the cube in $\mu m = 1*10^6 * 1*10^6 * 1*10^6 = 1*10^{18}\,\mu m^3$. Therefore, $1\,cm^3 = 1*10^{12}\,\mu m^3$, etc. Note that $1\,m = 100\,cm$, but $1\,m^3$ does not equal $100\,cm^3$.

You need to become familiar and comfortable with this sort of SI unit conversion – use the problems at the end of this chapter to practice. It is quite easy to get confused when performing this sort of calculation. One way to avoid this is to work completely in SI units, e.g. metres rather than millimetres, kilograms rather than grams, etc. Calculations involving the fundamental SI units often lead to very small or very big numbers, so you need to be comfortable with using scientific notation, as described in Chapter 6. You also need to be aware that you may need to convert your answer back into the units asked for or customarily used, e.g. millilitres rather than litres, or minutes instead of seconds.

## 3.3. SI usage

The SI system is precise not only about the units themselves, but also as to how they are written. There is a good reason for this, to avoid ambiguity, but unfortunately many people either ignore or are ignorant of these rules.

1. Many SI units are eponymous, i.e. named after famous people who were prominent in early work done within the field in which the unit is used. A unit which is named after a person is written in lower case (newton, volt, pascal etc.) when written in full, but should start with a capital letter (N, V, Pa, etc.) when abbreviated.

2. Units written in abbreviated form are *never* pluralized. This is because 'kms' does not mean 'kilometres', but 'km $*$ s'.

3. Unit abbreviations (such as J, N, g, Pa) are *never* followed by a full-stop unless at the end of a sentence.

4. The unit or abbreviation (plus prefix if present) should be separated from the number to which it refers by a space.

5. Prefixes are not separated from the unit to which they refer by spaces.

6. When units are combined they are separated by spaces, e.g. the acceleration due to gravity is $9.8 \, \mathrm{m\,s^{-2}}$.

7. '$-1$' is shorthand for 'per', e.g. $6 \, \mathrm{mg\,L^{-1}}$ means 6 milligrams per litre, and $9.8 \, \mathrm{m\,s^{-2}}$ means 9.8 metres per second per second. Never use 'p' for per, e.g. 'kph' should be written as $\mathrm{km\,h^{-1}} = $ kilometres per hour. Neither should you use a slash '/' to mean per, since this could be misinterpreted as a division symbol.

8. Any unit may take only *one* prefix. For example, 'millimillimetre' should be written as 'micrometre' ($\mu$m).

## 3.4. Measuring energy

Physics is the science of matter and energy and of the interactions between them. As such, it is fundamental to life, and plays a major role in areas

Table 3.4   Derived SI units

| Concept | Algebraic formula | Dimensional formula | SI unit |
| --- | --- | --- | --- |
| Force | Mass * acceleration | $kg * m\,s^{-2}$ | newton (**N**) |
| Work | Force * distance | $N * m$ | joule (**J**) |
| Power | Work/time | $J/s$ | watt (**W**) |
| Weight | Mass * acceleration due to gravity ($g$) | $N * 9.8\,m^{-2}$ | **N** (kg) |
| Gravitational potential energy | Weight * height | $N * m$ | J |
| Kinetic energy | $0.5 * mass * speed^2$ | $0.5 * kg * m\,s^{-2}$ | J |
| Pressure | Force/area | $N/m^2$ | pascal (**Pa**) |
| Electrical potential | Current * resistance | $A * \Omega$ | volt (**V**) |
| Electrical resistance | Potential/current | $V/A$ | ohm ($\mathbf{\Omega}$) |
| Electrical conductivity | 1/resistance | $1/\Omega$ | siemen (**S**) (not s = second) |

such as biomaterials and bioengineering, imaging, bioelectronics and nanobiology. However, in this book it is only appropriate to provide a brief explanation of basic physical principles.

Energy is the capacity of a system for work (the transfer of energy from one system to another) or power (the rate at which work is done, i.e. the amount of work per unit time). Force is an action which maintains or alters the position of a body, or distorts it (Table 3.4). Because force has both magnitude and direction, it is a vector quantity (not simply scalar, like mass, which has no direction). The mass of an object is a measure of an object's resistance to changes in either the speed or direction. Weight and mass are not the same thing. The weight of an object is the force it exerts under a given gravitational force. Thus the weight of an astronaut is different on the moon than on the Earth, but the mass is constant. Weight is therefore properly measured in newtons and mass in kilograms. However, we usually write about 'a weight of so-many kg'. This is not so much an error, but a shorthand. What is being referred to is the force with which the earth attracts a body. The phrase '10 kg weight' refers to an object on which the earth's gravity exerts a force of $10\,kg * 9.8\,m\,s^{-2} = 98\,N$. In practice, we refer to weight in kilograms, which assumes that the measurement is made under the acceleration due to gravity at sea level.

There are five main forms of energy:

1. *Mechanical energy* is associated with motion. Energy stored in an object due to its position is called potential energy. Gravitational potential energy is dependent on an object's height above the surface of the Earth. Kinetic energy is the energy an object has because of its motion.

2. *Heat* is caused by the motion of the particles which make up matter. When particles collide or when surfaces are rubbed together, mechanical energy is converted into heat. This is known as friction (a force which resists the relative motion or tendency to relative motion of two bodies in contact).

3. *Electromagnetic energy* is the energy given off by moving electric charges. Examples include light and other forms of electromagnetic radiation, such as radiowaves and microwaves.

4. *Chemical energy* is the energy released or absorbed when bonds between atoms are formed or broken. An example of this is the stored chemical energy in the fuel which powers an engine. When the fuel is burned, stored chemical energy is converted to heat and mechanical energy, which in turn is converted to motion. Photosynthetic organisms such as green plants and some bacteria convert the electromagnetic energy of light into stored chemical energy by combining carbon dioxide with water to produce carbohydrates. All other organisms (and photosynthetic organisms in the dark) utilize this stored energy to provide the power for life.

5. *Nuclear energy* is the energy stored in the nuclei of atoms. When the nucleus of an atom splits (fission), or when nuclei collide at high speeds and fuse together (fusion), nuclear energy is released. Nuclear energy from the sun reaches the Earth in the form of electromagnetic radiation and is the ultimate source of energy for all forms of life.

While it is possible to change matter into energy and energy into matter, the sum of the two is always constant. Therefore, matter and energy are different forms of the same thing which can be converted from one to the other. Conversion of one form of energy to another is quite common, e.g. potential energy changing into kinetic energy; kinetic energy changing into potential energy; chemical energy into heat energy; heat energy into mechanical energy; nuclear energy into heat energy; and mechanical energy into electromagnetic energy.

## 3.5. Temperature

The SI unit of temperature is the **kelvin** (1/273.16 of the thermodynamic temperature of the triple point of water). The triple point of water is the temperature at which water exists in its three possible phases, solid, liquid and vapour (0.01°C); 0 K represents absolute zero, the temperature at which all molecular motion stops. Water freezes at a temperature of 273.16 K ($=0.01$°C) and boils at 373.16 K ($=100$°C). Note that we do not write 'degrees kelvin' ('°K') as with °C, simply 'K'.

$$K = 273 + °C \text{ and } °C = K - 273$$

*Examples*

$$25°C = 273 + 25 = 298 \text{ K}$$
$$298 \text{ K} = 298 - 273 = 25°C$$

## Problems (answers in Appendix 1)

### True or false?

**3.1.** The litre (L) is the SI unit of volume.

**3.2.** The hertz (Hz) is the SI unit of frequency.

**3.3.** The kilogram (kg) is the SI unit of weight.

**3.4.** The Celsius is the SI unit of temperature.

**3.5.** The newton (N) is the SI unit of power.

### Are the following units written correctly or wrongly?

**3.6.** 3 kg.

**3.7.** The acceleration due to gravity is $9.8 \, m \, s^{-2}$.

**3.8.** The power output of the human heart is about 5 Watts.

**3.9.** A virus particle is $25 \, \mu mm$ in diameter.

**3.10.** The output of a nuclear power station is 23 GW.

## Converting SI units

**3.11.** Write 15 mm as nm.

**3.12.** Write 3 Pa as μPa.

**3.13.** Write $14 * 10^9$ g as kg.

**3.14.** Write $1 \, m^3$ as $cm^3$.

**3.15.** Write 1200 pg as ng.

**3.16.** What is the equivalent of 100°C in K?

**3.17.** What is the equivalent of 274 K in °C?

**3.18.** Write $0.005 \, kg \, cm^{-3}$ as $g \, m^{-3}$.

**3.19.** Convert $15 \, m \, s^{-1}$ into $km \, h^{-1}$.

**3.20.** Convert $29 \, m^3$ into L.

## Energy

**3.21.** What force is exerted in a chair by a 70 kg person standing on it?

**3.22.** A contracting muscle fibre exerts a force of 1 pN and moves the anchor point of the fibre 1 nm. How much work does the fibre do?

**3.23.** If the muscle fibre contracts in 55 ms, what power does it exert?

**3.24.** What is the weight of a seashell with a mass of 48 g lying on a beach?

**3.25.** What is the gravitational potential energy of a 2.1 kg coconut hanging from a tree 3.4 m above a beach?

**3.26.** What is the kinetic energy of a 2.1 kg coconut which strikes the head of a sunbather at a speed of $3.3 \, m \, s^{-2}$?

**3.27.** An electrophoresis gel carrying a current of 38 mA has a resistance of $5.25 * 10^4$ W. Calculate the voltage across the gel.

**3.28.** An electrophoresis gel has a resistance of $4.33 * 10^4$ W and a potential of 1920 V across the gel. What current is passing through the gel?

**3.29.** An electrophoresis gel has a potential of 1733 V and conducts a current of 55 mA. Calculate the electrical resistance of the gel.

**3.30.** A water sample has a resistance of $6505\,\mathrm{W\,cm^{-1}}$. Calculate the conductivity of the sample.

# 4

# Molarities and Dilutions

**LEARNING OBJECTIVES:**

On completing this chapter, you should be able to:

- understand the concepts of molarity and concentrations;
- be able to calculate the molarity of solutions;
- be able to calculate dilutions of solutions and suspensions.

## 4.1. Avogadro's number

In the last chapter, we learned that the mole (from 'molecule') is the SI unit used to measure the amount of a substance (abbreviation = 'mol'). Technically, one mole is 'the amount of a substance which contains as many elementary entities as there are atoms in 0.012 kg of carbon-12'. This is a complicated definition, but in 1811 Amedeo Avogadro was the first person to clearly make the distinction between molecules and atoms. Avogadro suggested that equal volumes of all gases at the same temperature and pressure contain the same number of molecules, which is now known as Avogadro's Principle. However, it was not until long after Avogadro's death that the idea of the mole was introduced. The molecular weight in grams (1 mol) of any substance contains the same number of molecules. The number of molecules in 1 mol is now called Avogadro's number. Avogadro had no knowledge of moles, or of the number that was to bear his name, since this was never determined by Avogadro himself.

Avogadro's number is:

$$602\,213\,670\,000\,000\,000\,000\,000 \; (6.02 * 10^{23})$$

How big is Avogadro's number?

1. An Avogadro's number of soft drink cans would cover the entire surface of the earth to a depth of over 200 miles.

2. If we were able to count atoms at the rate of 10 million per second, it would take about 2 billion years to count the atoms in 1 mol.

3. An Avogadro's number of carbon-12 atoms (1 mol) weighs 12 g. An Avogadro's number of moles of carbon-12 weighs the same as the Earth.

## 4.2. Molecular weight

To have to consider the masses of individual molecules measured in grams would be very inconvenient, since this would involve extremely small numbers. Unfortunately, the only clearly defined unit in this area is the SI standard, the mole, and the derived unit of concentration, molarity. Molecular weight is the mass of 1 mol of a substance (in grams per mole). However, the term 'molecular mass' is used by some people to mean molecular weight (i.e. grams per mole) and by others to mean mass of a single molecule (in daltons, Da – the mass of a molecule relative to one-twelfth the mass of carbon-12 atom). Although numerically equal, molecular weight/mass (in grams per mole) and mass-of-a-molecule type molecular mass (in daltons) have different units. These problems are avoided by using 'molecular mass' only for the mass of individual molecules (in daltons) and 'molecular weight' or 'molar mass' for the mass of a mole of molecules (grams per mole). Relative molecular mass ($M_r$) is an attempt to avoid these problems by defining the apparent molecular weight of proteins etc. with reference to a set of standards. However, if $M_r$ is defined as molecular mass relative to one-twelfth the mass of carbon-12 atom ( = daltons), then $M_r$ is really in daltons – a measurement which refers to individual molecules, not moles. $M_r$ is thus an illusion since it has no properly defined units, e.g. '$M_r$ 53 000' (53 000 what?). Therefore, since there is no consensus in this area, in this book the term 'molecular weight' is used to refer to any or all of these quantities.

> ## IMPORTANT
> The *mole* [abbreviated as **mol**, *not* m (metre), or M (molarity)]
> is a measure of the *amount* of a substance.
> *Molarity* (abbreviated as **M**) is a measure of the *concentration*
> of a substance, i.e. the amount of substance (number of moles)
> per unit volume (litres).
> ## Do not confuse these.

## 4.3. Solutions

A solution is a homogeneous mixture where all the particles (the 'solute')
exist as individual molecules or ions dissolved in a liquid (the 'solvent').
The molarity of a solution is calculated by dividing the moles of solute by
the volume of the solution (in litres):

$$\text{Molarity} = \frac{\text{moles of solute}}{\text{litres of solution}}$$

You may possibly encounter another term, the molality (abbreviated as
**m**) of a solution. Molality is the concentration of a solute measured as
moles per kilogram of solvent, cf. molarity, which is measured as moles of
solute per litre of solvent. For example, a 1 m NaCl solution contains
1 mol of NaCl per kilogram of water whereas a 1 M NaCl solution contains 1 mol of NaCl per litre of water. This is a small difference, but
molalities are preferred over molarities in experiments that involve temperature changes of solutions, e.g. calorimetry and freezing point depres-
sion experiments, where the volume of the solution changes. Molality is
almost never used in biology, but you should be aware that it exists and is
distinct from molarity, and is not just a typing mistake.

*Examples*

1. What is the molarity of 1 L of a solution containing 2 mol of solute?

$$\text{Molarity (M)} = \frac{2\,\text{mol}}{1\,\text{L}} = \mathbf{2\,M}$$

   i.e. $2\,\text{mol}\,\text{L}^{-1}$, or '2 molar' (2 M).

2. What is the molarity of 2.5 L of a solution containing 0.75 mol of solute?

$$\text{Molarity (M)} = \frac{0.75\,\text{mol}}{2.5\,\text{L}} = \mathbf{0.3\,M}$$

3. What is the molarity of 2 L of a solution containing 40 g of NaOH? This calculation must be performed in two stages:

- the molecular weight of NaOH is $40\,\text{g}\,\text{mol}^{-1}$. Number of moles is

$$\frac{40\,\text{g}}{40\,\text{g}\,\text{mol}^{-1}} = 1\,\text{mol}$$

- calculate molarity

$$\frac{1\,\text{mol}}{2\,\text{L}} = \mathbf{0.5\,M}$$

4. When 100 mL of a solution contain 2 g of NaCl (molecular weight $58.44\,\text{g}\,\text{mol}^{-1}$), what is the molarity of the solution?

$$M * V = \text{g/molecular weight}$$

$$M * 0.1\,\text{L} = \frac{2\,\text{g}}{58.44}$$

$$M * 0.1 = 0.034 = \mathbf{0.34\,M}$$

5. How many grams of NaCl are needed to make 500 mL of a 0.2 M solution?

$$M * V = \text{g/molecular weight}$$

$$(0.2\,\text{M}) * (0.5\,\text{L}) = \frac{x}{58.44}$$

$$0.1 = \frac{x}{58.44} = \mathbf{5.844\,g}$$

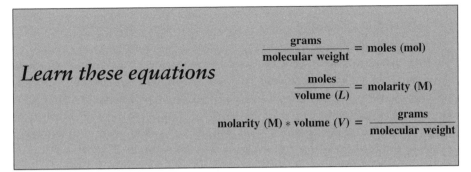

*Learn these equations*

$$\frac{grams}{molecular\ weight} = moles\ (mol)$$

$$\frac{moles}{volume\ (L)} = molarity\ (M)$$

$$molarity\ (M) * volume\ (V) = \frac{grams}{molecular\ weight}$$

If you are still not confident about molarity calculations, an alternative approach is to use a formula triangle:

## Examples

1. What is the concentration of 2 L of a solution containing 4 mol of solute?

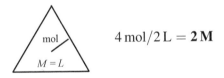

$$4\,mol/2\,L = \mathbf{2\,M}$$

2. How many moles of solute are contained in 2 L of a 0.5 M solution?

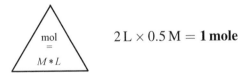

$$2\,L \times 0.5\,M = \mathbf{1\,mole}$$

3. What volume of 0.5 M solution can you make with 4 mol of solute?

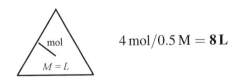

$$4\,mol/0.5\,M = \mathbf{8\,L}$$

## 4.4. Spectroscopy

Spectroscopy is a useful way of experimentally determining the concentration of biological molecules. Molecules in solution absorb light maximally at particular wavelengths. The absorption spectrum of a substance can be determined by measuring the optical density of a solution of the molecule over a range of wavelengths and finding the peak absorbance.

$$A = \varepsilon c l$$

where $A$ = the light absorbance of the solution at a defined wavelength; $\varepsilon$ = the molar extinction coefficient of the molecule, i.e. the ability of the molecule to absorb light of a specified frequency (sometimes known as the absorption coefficient); $c$ = the concentration of the molecule in the solution; $l$ = the length of the light path, i.e. how far the light travels through the solution – the width of the spectrophotometer cell (cuvette).

If we know three of these parameters it is easy to do some simple algebra to find the fourth, for example, if we measure the absorbance of a solution of known concentration in a cuvette with a known width (i.e. light path), we can calculate the extinction coefficient:

$$\varepsilon = A/(c * l)$$

More importantly, for a molecule with a known extinction coefficient, by measuring the light absorbance we can determine the concentration:

$$c = A/(\varepsilon * l)$$

This is the basis of many biological experiments, in particular enzyme assays where the appearance or disappearance of a particular reactant can be measured. One particularly useful aspect of this technology is that if the absorption spectra of different molecules in solution do not overlap significantly [e.g. enzyme, substrate and product(s)], we can accurately measure the concentration of one molecule without interference from the others.

### Example

The enzyme lactate dehydrogenase catalyses the following reaction:

$$\text{Pyruvate} + \text{NADH} + \text{H}^+ \rightarrow \text{lactate} + \text{NAD}^+$$

NADH [the reduced (electron-energy rich) coenzyme nicotinamide adenine dinucleotide] absorbs light with a wavelength of 340 nm with an extinction coefficient of $6.22 * 10^3 \, M^{-1} cm^{-1}$, whereas the oxidized form of the coenzyme, NAD, does not. By determining the rate at which NADH disappears, you can measure the rate of the reaction as lactate dehydrogenase converts NADH to NAD. Measuring the absorbance of the reaction mixture at 340 nm as a function of time will allow the reaction rate to be calculated. On adding the enzyme to a 3 mL reaction mixture in a cuvette with a 1 cm light path, the following measurements were made.

initial $A_{340}$: 0.78

$A_{340}$ after 1 min: 0.38

Initial concentration of NADH:

$$c = A/(\varepsilon * l)$$

$$c = 0.78/6.22 * 10^3 * 1 = 0.000125 \, M = 125 \, \mu M$$

$$125 \, \mu M = 125 \, \mu mol \, L^{-1}$$

Therefore the amount of NADH:

$$= 0.000125 \, M * (3 \, mL/1000 \, mL)$$
$$= 3.75 * 10^{-7} \, mol = 0.375 \, \mu mol$$

Concentration after 1 min:

$$c = 0.38/6.22 * 10^3 * 1 = 0.000061 \, M = 61 \, \mu M$$

$$61 \, \mu M = 61 \, \mu mol \, L^{-1}$$

$$Amount = 0.000061 \, M * (3 \, mL/1000 \, mL)$$

$$= 1.8 * 10^{-7} \, mol = 0.18 \, \mu mol$$

The amount of NADH used up in the reaction in 1 min is:

$0.375 \, \mu mol$ (initial amount) $- 0.18 \, \mu mol$ (amount remaining) $= 0.195 \, \mu mol$

Some molecules do not absorb visible light efficiently but do absorb shorter wavelengths such as ultraviolet (UV) light. The heterocyclic ring structures in DNA and RNA absorb light with a maximum absorbance at 260 nm. This can be measured by UV spectroscopy, using quartz cuvettes which transmit UV light. The extinction coefficient of DNA at 260 nm is approximately $10\,000 \, M^{-1} cm^{-1}$, which means that a solution of DNA with a concentration of $50 \, \mu g \, mL^{-1}$ has an $A_{260}$ of 1.0, while RNA has an extinction coefficient at 260 nm of approximately $12\,500\,M^{-1}\,cm^{-1}$, and a solution of RNA with a concentration of $40\,\mu g\,mL^{-1}$ has an $A_{260}$ of 1.0.

## 4.5. Dilutions

In the laboratory, dilutions of concentrated stock solutions are frequently used to make solutions of any desired molarity. To 'dilute' a solution means to add more solvent without the addition of more solute. (The resulting solution must be thoroughly mixed to ensure homogeneity.) The fact that the amount of solute remains constant allows calculations to be made:

moles of solute before dilution = moles of solute after dilution

From the definition of molarity (above):

moles of solute = molarity * volume

so we can substitute M * V (molarity * volume) into the above equation:

$$\textit{Learn this equation} \quad \mathbf{M_1 * V_1 = M_2 * V_2}$$

Volumes need not be converted to litres – any volume measurement is fine, so long as the same measurement is used on each side.

### Examples

1. You have 53 mL of a 1.5 M solution of NaCl, but need a 0.8 M solution. How many mL of 0.8 M NaCl can you make?

$$M_1 * V_1 = M_2 * V_2$$

$$1.5 * 53 = 0.8 * V_2$$

$$V_2 = 79.5/0.8 = \mathbf{99.38\,mL}$$

2. You need 225 mL of 0.6 M NaOH solution and you have a 2.5 M stock solution. How would you make up the solution?

$$M_1 * V_1 = M_2 * V_2$$

$$2.5 * V_1 = 0.6 * 225$$

$$V_1 = 135/2.5 = \textbf{54 mL}$$

If you are still not confident about dilutions an alternative approach is to use the *ratio method* to work them out:

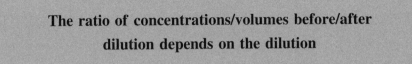

**The ratio of concentrations/volumes before/after dilution depends on the dilution**

## Examples

1. Fifty millilitres of a solution are diluted to a volume of 100 mL. The concentration of the diluted solution is 2 M. What was the concentration of the original solution?

$$50\,mL \rightarrow 100\,mL = \text{2-fold dilution}$$

Concentration of original solution (what you are trying to calculate) is *greater* than concentration of diluted solution, so *multiply* by the ratio of the volumes:

$$2\,M * 2 = \textbf{4 M}$$

2. What volume of a 5 M solution is required to make 100 mL of a 2 M solution?

$$5\,M \rightarrow 2\,M = \text{2.5-fold dilution}$$

The volume of the original solution (what you are trying to calculate) is *less* than the volume of the diluted solution, so *divide* by the ratio of the concentrations:

$$100/2.5 = \textbf{40 mL}$$

3. Five-hundred millilitres of a 5 M solution are diluted to 2 L. What is the concentration of the resulting solution?

$$0.5\,L \rightarrow 2\,L = 4\text{-fold dilution}$$

The concentration of resulting solution (what you are trying to calculate) is *less* than the concentration of the original solution, so *divide* by the ratio of the volumes:

$$5\,M/4 = \mathbf{1.25\,M}$$

4. Two-hundred millilitres of a 1 M solution are diluted to make a 0.4 M solution. What is the volume of the resulting solution?

$$1\,M \rightarrow 0.4\,M = 2.5\text{-fold dilution}$$

The volume of the resulting solution (what you are trying to calculate) is *greater* than the volume of the original solution, so *multiply* by the ratio of the volumes:

$$200 * 2.5 = \mathbf{500\,mL}$$

Frequently, the concentrations of solutions or suspensions are quoted not in molarities but in other terms such as weight per volume (e.g. $10\,g\,L^{-1}$ or $5\,mg\,mL^{-1}$), or as percentages. Weight percentage ('w/w' – the weight of solute multiplied by 100 divided by the weight of the solution) and weight–volume percentage ('w/v' – the weight of the solute in grams multiplied by 100 divided by the volume of solution in millilitres) express concentrations as percentages (parts per hundred). Aqueous reagents such as acids are often labelled in concentrations of weight percentage. Weight–volume percentage is often used for solutions made from solid reagents, e.g. a 15% sodium chloride solution contains 15 g of NaCl per 100 mL of solution, that is $150\,g\,L^{-1}$. Solutions of aqueous reagents are often described in terms of volume percentage ('v/v' – the volume of the solute in millilitres multiplied by 100 divided by the volume of solution in millilitres), e.g. 40% v/v ethanol contains 40 mL of ethanol per 100 mL of solution, that is 400 mL per litre.

Dilutions are frequently used not only with dissolved chemicals, but also with homogeneous suspensions of particles. Bacterial cultures grow to high densities of cells (up to $10^{10}$ cells per mL) and viruses multiply to high titres (up to $10^{12}$ infectious particles per mL). Bacterial densities are determined by counting colonies and virus titres determined by counting plaques in a biological assay. In order to obtain a countable number of plaques or colonies from a densely populated culture, it is usually necessary to dilute a sample of the culture.

A 1:1000 (1/1000) dilution could be prepared by adding 0.001 mL of culture (or solution) to 0.999 mL of diluent (solvent). However, these quantities are difficult to measure with precision and attempting to perform a large dilution such as this in a single step will result in inaccuracy. Even small inaccuracies in measuring the volumes involved in dilutions results in substantial error due to the fact that the multiplication involved in calculating the original concentration of the solution magnifies the inaccuracy.

Accuracy is achieved by performing a serial dilution – a series of small, accurate dilutions rather than a single large dilution.

## Example

To perform a 1:1000 (1/1000) dilution:

1. Prepare three tubes containing 0.9 mL of solvent/diluent.

2. Label the tubes: $10^{-1}$, $10^{-2}$ and $10^{-3}$.

3. Add 0.1 mL of the solution/culture to the $10^{-1}$ tube and mix thoroughly.

4. Transfer 0.1 mL from the $10^{-1}$ tube to the $10^{-2}$ tube and mix thoroughly.

5. Transfer 0.1 mL from the $10^{-2}$ tube to the $10^{-3}$ tube and mix thoroughly.

6. Perform the relevant assay, e.g. bacterial colony count, virus plaque assay, determine the optical density, etc., on the dilutions.

7. Calculate the results and remember to multiply the answer obtained by the reciprocal of the dilution to determine the concentration of the original culture or solution:

$$N = \frac{R * D}{V}$$

where $N$ = concentration of the original culture/solution; $R$ = result obtained from the assay; $D$ = reciprocal of the dilution tested; and $V$ = volume tested.

We could make this more accurate by making 10-fold dilutions by adding 1 mL culture or solution to 9 mL, since larger volumes can be measured with greater accuracy, although it is also more wasteful and you need to have enough of the original culture (1 mL). Alternatively, performing the dilution and assay in duplicate or triplicate and taking the average of the results obtained would also be more accurate, but possibly costly and certainly more laborious.

Note that adding 1 mL culture/solution to 10 mL solvent/diluent gives an 11-fold dilution, not a 10-fold dilution (since $1 + 9 = 10$ but $1 + 10 = 11$). If you attempted to perform a 1:1000 serial dilution like this, you would actually perform a $11 * 11 * 11 = 1{:}1331$ dilution.

Use the problems below to practise molarity and dilution concentrations.

# Problems (answers in Appendix 1)

4.1. One-hundred and twenty grams of NaOH are dissolved in water to make 5440 mL of solution. The molecular weight of NaOH is 40. What is the molarity of the resulting solution?

4.2. How many grams of NaCl are needed to make 120 mL of a 0.75 M solution? The molecular weight of NaCl is 58.44.

4.3. Sea water contains roughly 28.0 g of NaCl per litre. The molecular weight of NaCl is 58.44. What is the molarity of NaCl in sea water?

4.4. One-hundred and twenty-seven grams of NaCl and 19.9 g of sodium azide ($NaN_3$) are dissolved in water to make a 55 mL solution. The molecular weight of NaCl is 58.44 and that of $NaN_3$ is 65.01. What is the molarity of $NaN_3$ in the solution?

4.5. Undiluted sulphuric acid ($H_2SO_4$, molecular weight 98.07) is a 98% (w/v) solution. What is the molarity of the undiluted solution?

4.6. The molecular weight of bovine serum albumin (BSA) is 66 200. How many moles of BSA are present in 15 mL of 50 mg mL$^{-1}$ BSA solution?

4.7. What volume of 0.9 M KCl is needed to make 225 mL of 0.11 M solution? The molecular weight of KCl is 74.55.

**4.8.** 'TE' is a frequently used buffer solution for DNA and contains: 10 mM Tris–HCl pH 7.5 and 1 mM EDTA. You have a 1 M stock solution of Tris–HCl pH 7.5 and a 0.5 M stock solution of EDTA. What volume of each stock solution do you need to make 333 mL of TE buffer?

**4.9.** Chloramphenicol is soluble in ethanol at $0.1\,g\,mL^{-1}$ but much less soluble in water. What volume of $0.1\,g\,mL^{-1}$ solution must be added to 100 mL of a bacterial culture to give a final concentration of $150\,\mu g\,mL^{-1}$?

**4.10.** Urea lysis buffer contains the following ingredients:

   9.9 g urea ($M_w$ 60.06)/100 mL

   22 g SDS ($M_w$ 288.4)/100 mL

   77 ml 5 M NaCl stock solution/100 mL

   2.5 ml 0.2 M EDTA stock solution/100 mL

   15 ml 1 M Tris–HCl pH 8.0 stock solution/100 mL

   What are the final molar concentrations of *each* of the components of this buffer?

**4.11.** A solution of DNA with a concentration of $50\,\mu g\,mL^{-1}$ has an $A_{260}$ of 1.0. What is the concentration of the following solutions:

   (a) $A_{260}$ 0.65

   (b) $A_{260}$ 0.31 after diluting 15-fold

**4.12.** A solution of RNA with a concentration of $40\,\mu g\,mL^{-1}$ has an $A_{260}$ of 1.0. What is the concentration of the following solutions:

   (a) $A_{260}$ 0.59

   (b) $A_{260}$ 0.48 after diluting 10-fold

**4.13.** The amino acid tyrosine has an extinction coefficient of $1405\,M^{-1}\,cm^{-1}$ at 274 nm. Calculate:

   (a) The $A_{274}$ of a 0.4 mM solution of tyrosine measured in a cuvette with a 1 cm light path.

   (b) The concentration of a solution of tyrosine with $A_{274}$ 0.865 after 12-fold dilution measured in a cuvette with a 1 cm light path.

**4.14.** A sample of a culture of bacteria is subjected to 10-fold serial dilution; 0.1 mL aliquots of the dilutions are grown on agar plates and the number of colonies counted:

| Dilution: | $10^{-1}$ | $10^{-2}$ | $10^{-3}$ | $10^{-4}$ | $10^{-5}$ | $10^{-6}$ |
|---|---|---|---|---|---|---|
| Number of colonies: | Too many to count | Too many to count | 249 | 24 | 2 | 0 |

Assuming that one cell gives rise to one colony, how many cells were there per mL of the original culture?

**4.15.** A virus suspension was serially diluted to perform a plaque assay:

Dilution A was 0.1 mL + 4.9 mL

Dilution B was 1 mL of A + 1 mL

Dilution C was 0.01 mL of B + 0.99 mL

Dilution D was 0.1 mL of C + 9.9 mL

(a) Calculate the final dilution.

(b) Calculate the reciprocal of the final dilution.

(c) If the original suspension contained 12 000 000 virus particles $mL^{-1}$ and 0.1 mL of each dilution was used in the plaque assay, which dilution should be used to determine the density of the original suspension?

(d) Work out a simpler way of obtaining the same final dilution.

# Areas and Volumes

**LEARNING OBJECTIVES:**

On completing this chapter, you should be able to:

- write the formulae for calculating the areas and volumes of two- and three-dimensional objects.
- calculate the areas and volumes of complex shapes.
- apply this knowledge to biological problems.

## 5.1. Geometry

Geometry is the branch of mathematics that deals with the properties of space and objects. It is one of the oldest branches of mathematics, named from Greek for 'Earth measurement'. Trigonometry is the branch of mathematics concerned with specific functions of angles and their application to calculations in geometry. Classically, geometry deals with simple, regular shapes (Tables 5.1 and 5.2).

## 5.2. Calculating areas and volumes

Calculations involving simple geometric shapes are usually straightforward (except for ellipsoids) – use the formulae in Tables 5.1 and 5.2. However, it is sometimes said that 'There are no straight lines in biology'. Most biological shapes are complex, but are either approximations of simple shapes or

**Table 5.1**  Equations for perimeter and area of two-dimensional shapes

| Two-dimensional shapes | Perimeter | Area |
|---|---|---|
| Square | $4x$ | $x^2$ |
| Rectangle | $2(x+y)$ | $x*y$ |
| Circle | $2\pi r$ or $\pi*d$ $(d=2r)$ | $\pi r^2$ |
| Ellipse | $\pi[1.5(x+y) - \sqrt{(x*y)}]$ | $\pi*x*y$ |
| Triangle | $x+y+z$ | $0.5(z*h)$ |
| Right-angled triangle | $x+y+z$ | $0.5(x*y)$ |

combinations of simple shapes. Therefore, to determine shapes and volumes in biology, it is necessary to devise a strategy by which to approach the problem. This is a complex area which is impossible to define absolutely in simple terms, but the following examples illustrate some possible approaches.

## Examples

1. A tissue is composed of cells which are roughly **spherical** in shape and about 45 μm in diameter. Calculate the total number of cells present in 1 cm$^3$ of this tissue.

Table 5.2  Equations for surface area and volume of three-dimensional shapes

| Three-dimensional objects | Surface area | Volume |
|---|---|---|
| Cube | $6x^2$ | $x^3$ |
| Cuboid | $2(x*y) + 2(x*z) + 2(y*z)$ | $x*y*z$ |
| Sphere | $4\pi r^2$ | $(4\pi r^3)/3$ or $4/3(\pi r^3)$ |
| Ellipsoid | No simple formula | $4\pi(r*x*y)/3$ |
| Cylinder | $2\pi rh + 2\pi r^2$ | $\pi r^2 h$ |
| Cone/pyramid | $0.5(p*h+b)$ | $b*h/3$ |

- Calculate the volume of a single cell in $\mu m^3$:

formula for the volume of a sphere = $\mathbf{4/3(\pi r^3)}$

volume of each cell = $4/3(\pi\, 22.5^3\,\mu m^3) = 47\,713\,\mu m^3$

- Calculate the number of $\mu m^3$ in $1 \, cm^3$ of tissue:

$$1 \, cm = 10\,000 \, \mu m$$
$$1 \, cm^3 = 10^4 * 10^4 * 10^4 = 10^{12} \, \mu m^3$$

- Calculate the number of cells in $1 \, cm^3$ of tissue:

$$10^{12}/47\,713 = 20\,958\,648 = 2.09 * 10^7 \, cells$$

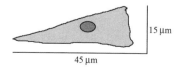

2. Fibroblast cells like the one illustrated above are cultured on circular glass coverslips 10 mm in diameter, and have a doubling time of 24 h. If you start with a coverslip with 10 000 cells on it, how long will it take for these cells to completely cover the coverslip?

- Convert everything into standard units, in this case $\mu m$.
- Calculate the area of the coverslip in $\mu m^2$.

$$\pi r^2 = 78.5 \, mm^2 = 7.85 * 10^7 \, \mu m^2$$

- Calculate the area of a single cell, assuming the cells are approximately triangular.

$$0.5(x * y) = 0.5 * 45 * 15 = 338 \, \mu m^2$$

- Calculate how many cells would cover the whole of the coverslip.

Area of coverslip/area of a single cell = number of cells required
$7.85 * 10^7/338 = 2.32 * 10^5$ cells required to cover the coverslip

- Calculate how long it will take for 1000 cells to multiply to give the required number.

By repeatedly multiplying 1000 by 2, it would take about 9 days for the original 10 000 cells to cover the coverslip.

| Day 1 | Day 2 | Day 3 | Day 4 | Day 5 | Day 6 | Day 7 | Day 8 | Day 9 |
|---|---|---|---|---|---|---|---|---|
| $1 * 10^3$ | $2 * 10^3$ | $4 * 10^3$ | $8 * 10^3$ | $1.6 * 10^4$ | $3.2 * 10^4$ | $6.4 * 10^4$ | $1.28 * 10^5$ | $2.56 * 10^5$ |

Are all the assumptions made valid? What practical aspects might cause inaccuracy in this situation?

- The calculation of the area of the coverslip is accurate.

- The cells are not exactly triangular and will be variable in shape, but in order to perform the calculation this is the best approximation that can be made.

- Calculating how many cells would cover the whole of the coverslip is quite accurate since the area of the coverslip is 'tessellated' into a mosaic of very small triangles.

- The last part of the calculation is probably the least accurate, since the rate at which the cells double is quite variable, and the growth rate will slow down as the coverslip becomes more crowded and the culture ages.

3. An *Arabidopsis* seedling has a root system which weighs 1.5 g. The average diameter of the roots is 0.2 mm. Calculate the total length and surface area of the root system.

Assume:

- the roots have the same density as water, i.e. $1 \, \mathrm{g \, mL^{-1}}$;

- each root is a cylinder with a constant radius;

- the root tips have a negligible volume.

Calculate the volume of the root system:

$$\mathrm{density} = \mathrm{weight/volume}$$

therefore

$$\mathrm{volume} = \mathrm{weight/density}$$

so

$$1.5 \, \mathrm{g} / 1 \, \mathrm{g \, mL^{-1}} = 1.5 \, \mathrm{mL}$$

Calculate the total root length:

$$\text{Volume of a cylinder} = V = \pi r^2 h$$

Therefore length of a cylinder $= h = V/\pi r^2$

$$h = 1.5/\pi * 0.01\,\text{cm}^2$$
$$= 47.8\,\text{cm} = 0.0478\,\text{m}$$

Calculate the surface area:

$$\text{Surface area of a cylinder} = 2\pi r h \quad \text{(ignoring the ends)}$$

$$2 * \pi * 0.01\,\text{cm} * 47.8\,\text{cm}$$
$$= 3\,\text{cm}^2$$

These are just a few examples of how the areas and volumes of complex biological shapes can be estimated. The art is to make the most valid assumptions which allow biological structures to be handled as if they were regular geometric shapes. There is usually more than one way of approaching such problems, and practising by completing the problems at the end of this chapter will allow you to improve your skills.

## Problems (answers in Appendix 1)

5.1. A snail travels in an elliptical path where the longest diameter is 1 m and the shortest diameter 0.5 m. How far does the snail travel?

5.2. The wing cases of a new species of beetle are shaped like right-angled triangles with sides of 1, 1.5 and 1.75 cm. What is the area of the wing cases? (Remember a beetle has two wing cases.)

5.3. An aquarium has internal dimensions of $109 * 47 * 47$ cm. What is its volume?

5.4. A conical antheap has a base area of 0.65 m$^2$ and a height of 0.24 m. What volume does the antheap occupy?

5.5. An onion root tip has a circumference of 19.9 μm at its base and a length of 9.2 μm (the tip has not been cut off from the plant, so you may assume it is a baseless cone). What are the surface area and volume of the root tip?

5.6. Lymphocytes (white blood cells) are essentially spherical in shape. In a blood sample, there are $2 * 10^6$ lymphocytes mL$^{-1}$. If the average diameter of a lymphocyte is 7.5 μm, what is the total surface area and total volume of all the lymphocytes in 1 mL of blood?

5.7. A waterlily pad is 10.5 cm in diameter and has an average thickness of 1.5 mm. What volume does it occupy and what is its total surface area? (You may assume the lilypad is a complete circle.)

5.8. A 20 μL blood sample from a mouse contains $9 * 10^4$ erythrocytes (red blood cells). If the mouse contains a total of 2.5 mL of blood and the average volume of an erythrocyte is $9 * 10^{-11}$ L, calculate:

   (a) the number of erythrocytes $mL^{-1}$ of blood;

   (b) the total number of erythrocytes in the mouse;

   (c) the total volume of the erythrocytes in the mouse.

5.9. The Earth is, on average, about $1.5 * 10^8$ km from the Sun. Earth's orbit is actually elliptical, and has different speeds at different parts of the orbit, but for this calculation, assume that the earth has a circular orbit.

   (a) How many metres does the Earth travel in one year?

   (b) How far does the Earth travel in one month?

5.10. The Earth rotates on its axis once every 24 h. A person at the equator travels through space in a (nearly) circular path with a radius of 6400 km. What is the person's speed?

# Exponents and Logs

**LEARNING OBJECTIVES:**

On completing this chapter, you should be able to:

- understand and be able to manipulate exponents;
- understand and be able to manipulate logarithms;
- be able to use logarithms to perform calculations;

## 6.1. Exponents

An exponent (also called the 'power' or 'index') of a number indicates how many times a number or term (the 'base') should be multiplied by itself. Just as multiplication is a shortcut for addition:

$$3 * 5 = 5 + 5 + 5$$

So exponents are a shortcut for multiplication:

$$5^3 = 5 * 5 * 5$$

Similarly,

$$5^9 = 5 * 5 * 5 * 5 * 5 * 5 * 5 * 5 * 5$$

and

$$(-5) * (-5) * (-5) = (-5)^3$$

and

$$(x + y) * (x + y) * (x + y) * (x + y) * (x + y) = (x + y)^5$$

As you will see later, logarithms are a shortcut for exponents, since the log function is the inverse of the exponential. How to manipulate and use exponents can be summarized in three rules.

1. Rule 1 – to **multiply** identical bases, **add** the exponents, e.g.

$$3^7 * 3^9 = (3 * 3 * 3 * 3 * 3 * 3 * 3) * (3 * 3 * 3 * 3 * 3 * 3 * 3 * 3 * 3)$$
$$= 3^{16}$$

2. Rule 2 – to **divide** identical bases, **subtract** the exponents, e.g.

$$\frac{4^9}{4^3} = \frac{4 * 4 * 4 * 4 * 4 * 4 * 4 * 4 * 4}{4 * 4 * 4}$$
$$= \frac{4}{4} * \frac{4}{4} * \frac{4}{4} * 4 * 4 * 4 * 4 * 4 * 4$$
$$= \frac{4^9}{4^3} = 4^{9-3} = 4^6$$

3. Rule 3 – when there are two or more exponents and only one base, multiply the exponents, e.g.

$$(3^4)^5 \text{ can be written as } 3^4 * 3^4 * 3^4 * 3^4 * 3^4$$

$$= 3^{4+4+4+4+4} = 3^{20}$$
$$= 3^{4*5} = 3^{20}$$

And, simplify $[(9^9)^9]^9$:

$$= 9^{9*9*9} = 9^{729}$$

**Remember**, these rules apply to identical bases only – you cannot apply them to different bases.

Note that zero raised to any power always equals zero ($0^n = 0$) and that any number raised to the power zero equals one ($n^0 = 1$). Negative exponents are the inverse of numbers raised to a positive integer:

$$a^{-n} = 1/a^n$$
$$\text{e.g. } 6^{-3} = 1/6^3 = 1/216 = 0.0046$$

Whenever you see a negative exponent, this should immediately suggest that the expression has a value of less than 1, e.g.

$$2^{-3} = 1/(2 * 2 * 2) = 1/8 = 0.125$$

Working in powers of 10:

$$10^3 = 1/10^{-3} = 1000$$
$$10^2 = 1/10^{-2} = 100$$
$$10^1 = 1/10^{-1} = 10$$
$$10^0 = 1/10^0 = 1$$
$$10^{-1} = 1/10^1 = 0.1$$
$$10^{-2} = 1/10^2 = 0.01$$
$$10^{-3} = 1/10^3 = 0.001$$

Fractional exponents can be dealt with in exactly the same way as integer exponents:

The square root of any number $= n^{0.5}$, e.g.

$$9^{0.5} = 3$$

and the cube root of any number $= n^{1/3}$, e.g.

$$8^{0.33} = 2$$

Using the rules of exponents:

$$(2^5)^{0.2} = 2^{(5*0.2)} = 2^1 = 2$$

In real life no one expects you to work out complex exponents by hand – use a scientific calculator. To calculate the value of $10^{0.65}$ type this expression into your calculator and you will see that it equals 4.47. However, the reason for the explanations in this chapter is that, even with a calculator, you still need to understand how exponents work. So what has all this got to do with the real world? Here is a practical example you may be interested in. If you invested £1200 in a bank account which pays 6% interest compounded annually, how much money would be in your account after 5 years?

If the annual interest rate is 6 per cent, this can be written as 0.06.

If you started with £1, at the end of one year, you would have $1 + 0.06 = £1.06$.

At the end of the second year, you would have £1.06 + 0.06 * $(1 + 0.06) = (1.06) * (1.06) = (1.06)^2$.

At the end of 5 years, you would have $(1.06)^5 = £1.33$ for each pound invested, so:

$$1.33 * 1200 = £1596$$

## 6.2. Exponential functions

The expression $y = z^x$ solved for $y$ in terms of $x$ does not graph as a straight line on normal graph paper (Figure 6.1). Any quantity which increases by being multiplied by the same value at regular intervals is said to grow 'exponentially', i.e. each subsequent value is equal to the previous value multiplied by a constant, $z$:

$$z^0 = 1$$
$$z^1 = z^0 * z = z$$
$$z^2 = z^1 * z = z * z \text{ etc.}$$

Exponential functions occur frequently in biology because they describe processes of growth and decay, for example in radioactive decay or bacterial growth.

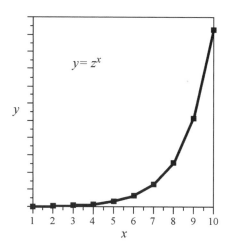

**Figure 6.1**  An exponential graph

## Radioactive decay

Radioisotopes are unstable forms of elements, atoms of which decay spontaneously at random. Radioactive decay is described by the function:

$$N = N_0 e^{-\lambda t}$$

where $N$ is the amount of radioactivity remaining at time $t$, $N_0$ the original amount of radioactivity and $\lambda$ the decay constant for the particular radionuclide. Since the speed at which the element decays is constant, this is an exponential function. The term 'half-life' is used to describe the time taken for 50 per cent of the atoms in a sample to decay. The half-life of different radionuclides varies widely, from 300 000 years in the case of $^{36}$Cl to a few seconds in other cases. Living organisms contain a mixture of radioactive $^{14}$C (half-life 5730 years) and the stable isotope of carbon, $^{12}$C. Once an organism dies the $^{14}$C is no longer replaced and slowly decays according to the exponential decay law:

$$N = N_0 e^{-\lambda t}$$

where $N$ is the amount of radioactivity remaining after time $t$; $N_0$ is the original amount of radioactivity; $\lambda$ is the decay rate (as a decimal); and $t$ the elapsed time.

Thus the age of any object which once contained living tissue can be estimated from the amount of $^{14}$C remaining.

## Bacterial growth

During the early phase of growth in a bacterial culture, before nutrients are exhausted or toxins build up, the number of cells doubles at regular intervals as the cells divide:

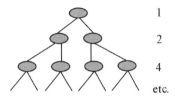

1

2

4

etc.

This phase of 'exponential growth' can be described by an exponential function:

$$N(t) = N_0 2^{t/t_0}$$

where $N$ = the number of cells at time $t$ and $N_0$ = the number of cells at time zero ($t_0$). The growth of populations is discussed in further detail below.

## 6.3. Logarithms

The term 'logarithm' comes from the Greek words *logos*, meaning 'to calculate', and *arithmos*, 'a number'. Just as exponents are a shortcut for multiplication, logarithms are a shortcut for exponents. A logarithm is the exponent or power to which a base must be raised to yield a given number (Table 6.1).

Logarithms are written as the sum of an integer (the 'characteristic') and a decimal (the 'mantissa', e.g. $\log_{10} 150 = 2.176$: characteristic = 2, mantissa = 0.176). Logarithms were invented by John Napier in 1614. Logarithms make it possible to do multiplication and division simply by adding and subtracting:

1. To **multiply** two numbers, **add** their logs, then determine the antilog of the result from tables or using a calculator.

2. To **divide** two numbers, **subtract** their logs, then determine the antilog of the result from tables or using a calculator.

Logs are also useful for many other calculations, for example to work out the fifth root of a number, divide the logarithm of the number by 5:

$$\text{The fifth root of } 10\,000\,000\,000 = \log_{10} 10\,000\,000\,000 = 10$$

Divide by 5:

$$10/5 = 2$$
$$\text{antilog } 2 = 100$$

Table 6.1  Logarithms

| Base | Exponent | Resulting number | Log |
|------|----------|------------------|-----|
| 10 | 1, i.e. $10^1$ | 10 | 1 |
| 10 | 2, i.e. $10^2$ | 100 | 2 |
| 10 | 3, i.e. $10^3$ | 1000 | 3 |
| 10 | 4, i.e. $10^4$ | 10 000 | 4 |

Logarithms can be converted back to real numbers by looking up tables of antilogarithms, the inverse of the logarithm function (e.g. $\log x = 2.176$, $x = $ antilog 2.176), or on a calculator as follows:

- either the $10^x$ key with $x = $ logarithm value, e.g. 2.176;
- or the INV key then the log key;
- or the $y^x$ key with $y = 10$ and $x = $ logarithm value.

The most commonly encountered logarithms are in base 10, written as '$\log_{10}$'. Logarithms to the base 'e' (2.71828...) are known as natural logarithms and are written as 'ln' (e is Euler's number, named after a Swiss mathematician who described e in 1728). As with exponents, there are some simple rules which can help you work with logarithms. Since logarithms are nothing more than exponents, these rules come from the rules of exponents (above):

There are three rules that apply under conditions where: $a$ is a positive number not equal to 1; $n$ is a real number (i.e. all numbers representable by an infinite decimal expansion, that is the ratio $a/b$, where $a$ and $b$ are integers and $b$ does not equal zero); and $x$ and $y$ are positive real numbers.

1. Rule 1 – $\log_a(x * y) = \log_a(x) + \log_a(y)$, i.e. you can multiply two numbers by adding their logarithms.

2. Rule 2 – $\log_a(x/y) = \log_a(x) - \log_a(y)$, i.e. you can divide two numbers by subtracting their logarithms.

3. Rule 3 – $\log_a(x)^n = n\log_a(x)$, i.e. the logarithm of an exponent gives the original number.

Remember the exponential function $y = z^x$ described in Section 6.2? When this function is plotted on semi-log graph paper (also called log–linear graph paper, where the $y$ axis has a log rather than a linear scale), or if the log values of $y$ are plotted, the log transformation of the exponential function results in the exponential curve being transformed into a straight line (Figure 6.2).

Logarithms can be used to solve exponential equations which occur commonly in biology, e.g. pH:

$$pH = -\log[H^+]$$

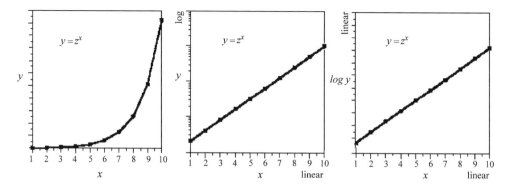

**Figure 6.2**  The log transformation of the exponential function results in the exponential curve being transformed into a straight line

Human blood plasma has a typical $H^+$ concentration (written as '$[H^+]$') of $10^{-7.4}$ M. Therefore:

$$\text{pH of blood} = -\log 10^{-7.4} = 7.4$$

Another example is the exponential growth of populations. The growth of a population is described by the equation:

$$N(t) = N_0 e^{\lambda t}$$

where $N$ is the size of the population at time $t$; $N_0$ is the size of the original population; e is Euler's number; and $\lambda$ is the growth rate (as a decimal).

During the growth phase of a bacterial culture, the rate of increase of cells is proportional to the number of bacteria present. The constant of proportionality, $\mu$, is an index of the growth rate and is called the growth rate constant:

$$\text{Rate of increase of cells} = \mu * \text{number of cells}$$

The value of $\mu$ can be determined from the following equation:

$$\ln N_t - \ln N_0 = \mu(t - t_0)$$

The natural logarithm of the number of cells at time $t$ minus the natural logarithm of the number of cells at time zero $(t_0)$ equals the growth rate constant multiplied by the time interval. For most purposes, it is easier to use $\log_{10}$ values rather than natural logarithms. Since natural logarithms are to the base e, the above equation can be converted to $\log_{10}$ by dividing by 2.303 (note: $\ln x = 2.303 \log x$, so the natural log of any number divided by 2.303 gives the $\log_{10}$ of that number):

$$\log_{10} N - \log_{10} N_0 = (\mu/2.303)(t - t_0)$$

or alternatively

$$\mu = [(\log_{10} N - \log_{10} N_0)2.303]/(t - t_0)$$

By measuring the increase in the number of cells during a certain time period, the growth rate constant ($\mu$) can be calculated, e.g. if:

$$t_0 = 1.5\,\text{h}, \quad N_0 = 8.4 * 10^1, \quad \text{then } \log_{10} N_0 = 1.92$$
$$t = 8.5\,\text{h}, \quad N = 3.39 * 10^8, \quad \text{and } \log_{10} N = 8.53$$
$$\mu = [(\log_{10} N - \log_{10} N_0)2.303]/(t - t_0)$$

Therefore in this case:

$$\mu = [(8.53 - 1.92)2.303]/(8.5 - 1.5)$$
$$\mu = 2.18\,\text{h}^{-1}$$

A further example is the decibel scale of sound intensity:

$$\text{dB} = 10 \log_{10} I/I_0$$

where $I$ is the intensity of the sound and $I_0$ is the intensity of a reference sound $(1 * 10^{-12}\,\text{Wm}^{-2}$, the lowest sound intensity detectable by the human ear). We use decibels because the ear is capable of hearing a very large range of sound – more than a billion-fold difference in intensity. To deal with such a range, logarithmic units are most useful.

If the sound intensity at a heavy metal concert is recorded as $0.5 * 10^{-1}\,\text{Wm}^{-2}$, what is the intensity of the sound in decibels?

$$\text{dB} = 10 \log_{10} (I/I_0)$$
$$= 10 * \log_{10}(0.5 * 10^{-1}/1 * 10^{-12})$$
$$= 10 * 10.7$$
$$= 107\,\text{dB}$$

The engine of a Ferrari makes a noise of 105 dB and a Ford engine makes a noise of 99 dB. How many times more intense is the Ferrari engine noise?

$$\text{difference in noise} = 105 - 99 = 6\,\text{dB}$$

Remember that a difference of 10 dB represents a 10-fold difference in sound intensity, so the Ferrari is six times louder than the Ford.

## Problems (answers in Appendix 1)

**6.1.** Simplify (calculate the value of):     $8^4 * 8^4$

**6.2.** Simplify:     $8^5 / 8^4$

**6.3.** Simplify:     $(8^5)^5$

**6.4.** Simplify:     $\log_{10}(5 * 4)$

**6.5.** Simplify:     $\log_{10}(5/4)$

**6.6.** Simplify:     $\log_{10}(3.3)^3$

**6.7.** What is the pH of a 0.011 M solution of HCl?

**6.8.** What is the pH of 100 mL of a solution containing 9 mg of HCl? The molecular weight of HCl is 36.46.

**6.9.** What is the $H^+$ concentration in a solution of HCl with a pH of 3?

**6.10.** In an exponentially growing bacterial culture: number of cells mL$^{-1}$ ($N_0$) at 3 p.m. ($t_0$) $= 5.5 * 10^3$; Number of cells mL$^{-1}$ ($N$) at 5 p.m. ($t$) $= 2.5 * 10^6$. Calculate:

1. $\log_{10} N_0$

2. $\log_{10} N_t$

3. $t - t_0$

4. $\log_{10} N - \log_{10} N_0$

5. $\mu$

**6.11.** A skeleton recovered from a peat bog is examined by police forensic scientists to determine whether a crime has been committed. Based on the decay of $^{14}C$ (half-life 5730 years), they estimate that 49% of the original $^{14}C$ in the skeleton has decayed. Calculate the probable age of the skeleton.

**6.12.** The population of a wild mink on a river system increases from 2300 to 3245 in 7 years. What was the population at the end of the first year? How long will it take for the original population to double?

# 7

# Introduction to Statistics

---

**LEARNING OBJECTIVES:**

On completing this chapter, you should understand:

- statistical variables;
- how to construct a frequency distribution;
- how to calculate percentiles.

---

## 7.1. What is statistics?

Statistics is the systematic collection and display of numerical data. Unfortunately, statistics is also the most abused area of numeracy. The reason for this is that, although it is easy to generate a single number to represent a complex dataset, unless appropriate statistical techniques are used the results are meaningless – garbage in, garbage out!

Statistical analysis should ideally follow a defined procedure:

1. Research question.

2. Statistical question.

3. Data collection.

4. Statistical conclusion.

5. Research conclusion.

Frequently this is not the case, and you will be presented with data you have not been involved with collecting and asked to draw conclusions. Such a process contains many possible pitfalls and to avoid them it is necessary to have a firm understanding of different statistical methods and their limitations. Before we can do this, however, it may be a good idea to try to explain some of the jargon which surrounds statistics.

## 7.2. Statistical variables

Different classes of information are known as the variables of a dataset, e.g. age, weight, height, gender, marital status, annual income, etc. Variables which are experimentally manipulated by an investigator in an experiment are called *independent variables*. Variables which are measured are called *dependent variables*, since they depend on the way in which the independent variables have been set up. All other factors which may affect the dependent variable(s) are called confounding, extraneous or secondary variables. These are important, since unless they are the same for each group being tested, comparisons between groups and statistical conclusions will be unreliable. Variables may be classified as either quantitative or qualitative (Figure 7.1).

*Qualitative data* provide labels, or names, for categories of like items, i.e. a set of observations where any single observation is a word or code that represents a class or category. Examples of qualitative variables are age (see below), gender and marital status. Qualitative data can be further divided into:

1. Nominal variables – variables with no inherent order or ranking sequence, e.g. numbers used as names (group 1, group 2 ...), gender, etc.

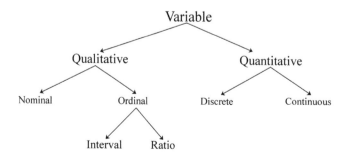

**Figure 7.1**  The different classes of variables

2. Ordinal variables – variables with an ordered series, e.g. 'greatly dislike, moderately dislike, indifferent, moderately like, greatly like'. If numbers are assigned to such variables they indicate rank order only – the 'distance' between the numbers has no inherent value.

3. Interval variables – equally spaced variables, e.g. temperature. The difference between a temperature of 66° and 67° is the same as the difference between 76° and 77°. Interval variables do not have a true zero, e.g. 88° is not necessarily 'double' the temperature of 44° this all depends on the scale (e.g. Celsius or kelvin).

4. Ratio variables – variables spaced at equal intervals with a true zero point, e.g. age.

Quantitative data measures either how much or how many of something, that is, a set of observations where any single observation is a number that represents an amount or a count. Examples of quantitative variables are weight, height and annual income. Quantitative data can be divided into:

1. Discrete variables – the set of all possible values which consists only of isolated points, e.g. counting variables (1, 2, 3 ...). An example of such data might be the number of students in a class, which will always be an integer, and never '53.7'.

2. Continuous variables – within the limits of the variable range, continuous variables can take on any possible value. Examples of such data might be the length or weight of a group of animals, e.g. 52.1, 52.8 and 52.9 g.

# 7.3. Statistical methods

Statistics is the science of collecting and displaying numerical data. Statistics are used to summarize or describe the basic features of datasets and present quantitative descriptions in a manageable form, for example, how do the marks on this year's course compare with last year's marks? Descriptive statistics (Chapter 8) are used to analyse the basic features of data under consideration. They do not draw conclusions from the data, but merely illustrate patterns or trends. There are many different descriptive methods which can be used for this purpose.

**Tabular methods** can be used to represent data sets with two or more variables, e.g.

| Gender | Age | | | | |
|--------|-------|-------|-------|-------|-------|
|        | 20–29 | 30–39 | 40–49 | 50–59 | 60–69 |
| Male   | 6     | 7     | 13    | 12    | 7     |
| Female | 12    | 14    | 10    | 9     | 11    |

**Numerical measures** of data sets are diverse, including (but not limited to): proportion, percentage, mean (average), median, mode, percentiles, range, variance and standard deviation. These are described in detail in the next chapter.

**Graphical methods** comprise all visual methods which summarize data, e.g.

1. Bar charts/histograms.

2. Pie charts.

3. Scatter diagrams.

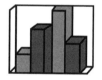

Graphs make it particularly easy to see features in numerical data. Unfortunately, as with other statistical methods, graphs are frequently abused. Some things you may have forgotten about graphs include:

1. Graphs generally have two axes (plural of axis), a horizontal $x$-axis (or abscissa) and a vertical $y$-axis (or ordinate), which should be clearly labelled.

2. The intersection between the axes represents the zero point – if it does not, the graph should make this clear, e.g. by axis labels or breaks.

3. The axes should be long enough so that values and labels can be easily read, but not hugely dissimilar in length.

4. A graph should be sufficiently clear that it does not need an accompanying explanation other than a brief title or legend. If any graph is not this clear, the data has been presented in the wrong format.

Chapter 8 contains a further explanation of descriptive statistics.

## 7.4. Frequency distributions

The basis of most statistical investigations is construction of a frequency distribution, the number of observations for each of the possible categories in a dataset. For nominal variables, the order in which the categories are listed makes no difference. For ordinal, interval, ratio variables and all quantitative data, the categories should be listed in rank order (e.g. Table 7.1).

### Grouped frequency distributions

Remember that the aim of descriptive statistics is to analyse the basic features of a dataset. Complex data can be simplified by combining individual scores to form a smaller number of groups, referred to as class intervals. These intervals should be: mutually exclusive, i.e.

**Table 7.1**  How to list categories in rank order

| Age | Frequency | | Age | Frequency |
|-----|-----------|-----|-----|-----------|
| 1 | 11 | | 5 | 2 |
| 2 | 12 | or | 4 | 6 |
| 3 | 19 | | 3 | 19 |
| 4 | 6 | | 2 | 12 |
| 5 | 2 | | 1 | 11 |

**Table 7.2** Class intervals

| Marks | Frequency | | Marks | Frequency |
|-------|-----------|------|-------|-----------|
| 91–100 | 1 | | 90–100 | 1 |
| 81–90 | 2 | *not* | 80–90 | 2 |
| 71–80 | 9 | | 70–80 | 9 |
| 61–70 | 0 | | 50–60 | 2 |
| 51–60 | 2 | | | |

non-overlapping; all of the same width; and continuous throughout the distribution (e.g. Table 7.2).

## Cumulative frequency distributions

Cumulative frequency distributions are useful to show what proportion of a dataset lies above or below certain limits (e.g. Table 7.3).

What percentage of this class scored the required pass mark of 41%? By looking at the cumulative percentage column, it is easy to see that 10% of the class scored 40% or less, so 90% achieved the pass mark.

Percentiles are points on a frequency distribution below which a specified percentage of cases in the distribution falls, for example a person scoring at the 75th percentile did better than 75% of those in the distribution:

$$\text{Percentile} = \frac{(n+1)P}{100}$$

where $n =$ number of cases and $P =$ desired percentile.

**Table 7.3** Cumulative frequency distribution

| Marks | Frequency | Cumulative frequency | Cumulative percentage |
|-------|-----------|----------------------|-----------------------|
| 91–100 | 1 | 50 | 100 |
| 81–90 | 9 | 49 | 98 |
| 71–80 | 9 | 40 | 80 |
| 61–70 | 13 | 31 | 62 |
| 51–60 | 7 | 18 | 36 |
| 41–50 | 6 | 11 | 22 |
| 31–40 | 4 | 5 | 10 |
| 21–30 | 0 | 1 | 2 |
| 11–20 | 0 | 1 | 2 |
| 1–10 | 1 | 1 | 2 |
| | $n = 50$ | | |

Table 7.4  Quartiles

| Marks | Cumulative frequency | Cumulative percentage | Percentile |
| --- | --- | --- | --- |
| 99 | 15 | 100 | |
| 93 | 14 | 93 | |
| 86 | 13 | 87 | |
| 80 | 12 | 80 | $P_{75}$ |
| 74 | 11 | 73 | |
| 68 | 10 | 67 | |
| 61 | 9 | 60 | $P_{50}$ |
| 55 | 8 | 53 | |
| 49 | 7 | 47 | |
| 42 | 6 | 40 | |
| 36 | 5 | 33 | $P_{25}$ |
| 30 | 4 | 27 | |
| 24 | 3 | 20 | |
| 17 | 2 | 13 | |
| 11 | 1 | 7 | |

75% of dataset

50% of dataset

25% of dataset

The 25th, 50th and 75th percentiles are also referred to as quartiles ($Q_1$, $Q_2$, $Q_3$), since they divide the dataset into quarters (e.g. Table 7.4). In this example, the range (from the highest datapoint to the lowest datapoint) is 88 (highest mark − lowest mark, i.e. $99 - 11 = 88$). Therefore, in this example, each quarter of the dataset is 22, so:

$$P_{25} = 11 + 22 = 33$$
$$P_{50} = 33 + 22 = 55$$
$$P_{75} = 55 + 22 = 77$$

Calculating percentiles from grouped data is slightly more complicated. An interpolation method is required:

1. Find the group within which the percentile lies.

2. Determine the percentage between the bottom of the distribution and the group containing the percentile.

3. Determine the number of additional datapoints required to make up the percentile.

**Table 7.5**  Calculating a percentile

| Marks | Frequency | Cumulative frequency | Cumulative percentage | Examples |
|-------|-----------|----------------------|-----------------------|----------|
| 91–100 | 1 | 50 | 100 | |
| 81–90 | 9 | 49 | 98 | $P_{90}$: $(k*n/100) = 90*50/100 = 45$ |
| 71–80 | 9 | 40 | 80 | |
| 61–70 | 13 | 31 | 61 | $81 + [(45-40)/9]*10 = 87\%$ |
| 51–60 | 7 | 18 | 36 | |
| 41–50 | 6 | 11 | 22 | $P_{20}$: $(k*n/100) = 20*50/100 = 10$ |
| 31–40 | 4 | 5 | 10 | |
| 21–30 | 0 | 1 | 2 | $41 + [(10-5)/6]*10 = 49\%$ |
| 11–20 | 0 | 1 | 2 | |
| 1–10 | 1 | 1 | 2 | |
| | $n = 50$ | | | |

4. Assume that the scores in the group are evenly distributed. (This assumption is necessary since there is no way to know the exact distribution within each group.)

5. Find the additional number of datapoints in the group required to make up the percentile.

6. Add this to the number of datapoints between the bottom of the distribution and the group containing the percentile.

$$\text{Percentile} = \text{LCB} + \{[(k*n/100) - \text{CFB}]/f\} * i$$

where LCB = the lower class boundary of the interval containing the percentile; $k$ = percentile; $n$ = total number of scores in the distribution; CFB = cumulative frequency of all the intervals below the interval containing the percentile; $f$ = the frequency (i.e. number of scores) in the interval containing the percentile; and $i$ = interval size (i.e. range of scores in the interval containing the percentile).

To find a percentile, $(k*n/100)$ must be calculated first in order to identify the interval in which the percentile is located (e.g. Table 7.5).

## 7.5. Frequency distribution graphs

Graphs can make it particularly easy to see features in numerical data. For example, a cumulative frequency curve reveals the shape of a frequency distribution (Figure 7.2).

| Marks | Frequency | Cumulative frequency | Cumulative percentage |
|-------|-----------|----------------------|------------------------|
| 90–99 | 0 | 50 | 100 |
| 80–89 | 2 | 50 | 100 |
| 70–79 | 6 | 48 | 96 |
| 60–69 | 9 | 42 | 84 |
| 50–59 | 12 | 33 | 66 |
| 40–49 | 10 | 21 | 42 |
| 30–39 | 7 | 11 | 22 |
| 20–29 | 3 | 4 | 8 |
| 10–19 | 1 | 1 | 2 |
| 0–9 | 0 | 0 | 0 |

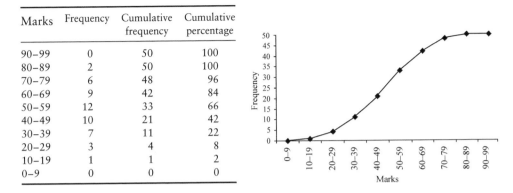

**Figure 7.2**  A cumulative frequency curve reveals the shape of a frequency distribution

| Marks | Frequency | Cumulative frequency | Cumulative percentage |
|-------|-----------|----------------------|------------------------|
| 90–99 | 0 | 50 | 100 |
| 80–89 | 2 | 50 | 100 |
| 70–79 | 6 | 48 | 96 |
| 60–69 | 9 | 42 | 84 |
| 50–59 | 12 | 33 | 66 |
| 40–49 | 10 | 21 | 42 |
| 30–39 | 7 | 11 | 22 |
| 20–29 | 3 | 4 | 8 |
| 10–19 | 1 | 1 | 2 |
| 0–9 | 0 | 0 | 0 |

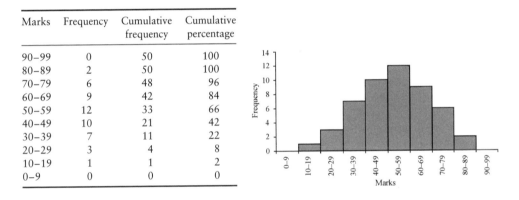

**Figure 7.3**  A frequency histogram

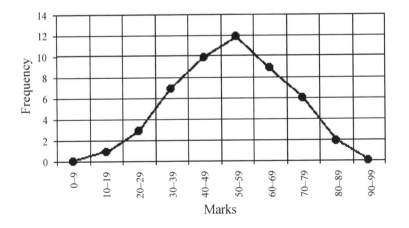

**Figure 7.4**  Frequency polygon

## Histograms

A frequency histogram is a series of rectangles representing the frequencies of the class intervals (Figure 7.3). The same data can also be drawn as a frequency polygon (Figure 7.4).

Which format is best? To some extent, the choice of graph formats is a matter of personal preference, but in general a histogram is best used for ungrouped data and discrete variables, and a polygon is best for grouped data and continuous variables. However, other graph formats can also be used. A bar diagram has gaps between the categories (groups) and therefore should only be used to represent discrete variables or categories

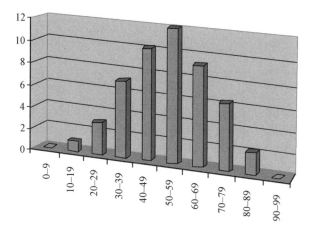

Figure 7.5   A bar diagram

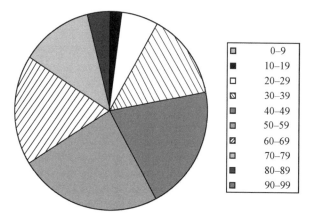

Figure 7.6   A pie diagram

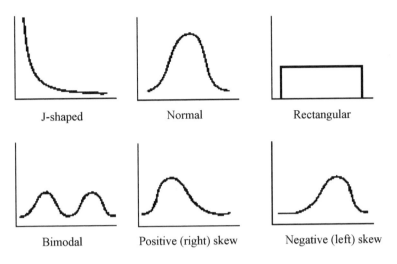

**Figure 7.7**   Different-shaped frequency distributions

(Figure 7.5). Pie diagrams represent relative frequencies, so can be used for grouped data and continuous variables (Figure 7.6).

Graphical methods are an important part of statistics, being perhaps the most powerful way of revealing the shape of data distributions (Figure 7.7). Different-shaped frequency distributions are described by their degree of skew (lack of symmetry about the mean – see next chapter) and kurtosis (the size of the 'tails', extremities, of the curve). This can also affect the calculation of some (but not all) statistics, as you will see in the next chapter.

## Problems (answers in Appendix 1)

**7.1.** Identify which type of variable each of the following parameters corresponds to:

(a)  blood type (A, B, AB, O);

(b)  number of eggs in a nest;

(c)  temperature – Celsius;

(d)  temperature – kelvin;

(e)  age;

(f)  questionnaire (e.g. terrible, poor, average, good, very good);

(g)  gender;

(h)  height;

(i) apple variety (e.g. Cox, Discovery, etc.);

(j) number of blood cells per mL.

**7.2.** Construct a grouped frequency distribution table for the following data showing the cumulative frequencies and cumulative percentage:

16, 33, 27, 82, 99, 14, 17, 74, 57, 83, 43, 27, 69, 82, 24, 25, 9, 2, 37, 85

**7.3.** Are the following frequency distribution tables constructed correctly or wrongly?

(a)

| Group | Frequency |
|-------|-----------|
| Group 4 | 29 |
| Group 2 | 24 |
| Group 3 | 22 |
| Group 1 | 17 |
| Group 5 | 9 |

(b)

| Scores | Frequency | Cumulative frequency |
|--------|-----------|----------------------|
| 1–10 | 6 | 6 |
| 11–20 | 2 | 8 |
| 21–30 | 14 | 22 |
| 41–50 | 27 | 49 |
| 61–70 | 33 | 82 |
| 71–80 | 17 | 99 |
| 81–90 | 14 | 113 |
| 91–100 | 2 | 115 |

(c)

| Scores | Frequency | Cumulative frequency |
|--------|-----------|----------------------|
| 3–9 | 1 | 1 |
| 10–16 | 3 | 4 |
| 17–24 | 5 | 9 |
| 25–30 | 9 | 18 |
| 31–38 | 4 | 22 |
| 39–45 | 2 | 24 |
| 46–53 | 7 | 31 |
| 54–61 | 11 | 42 |

**7.4.** For the following dataset:

16, 33, 65, 82, 99, 14, 17, 74, 57, 83, 43, 27, 69, 82, 24, 25, 9, 2, 37, 85, 1, 13, 96

(a) Calculate the scores at the first ($Q_1$) and third ($Q_3$) quartiles.

(b) Calculate the scores at the 45th ($P_{45}$) and 95th ($P_{95}$) percentiles.

**7.5.** Which of these graph types:

scatter plot, histogram, pie diagram

would be suitable for the following datasets?

(a) Proportion of university students from different schools;

(b) enzyme activity at a range of pH measurements;

(c) number of undergraduates, 1999–2005;

(d) blood pressure against time;

(e) grouped frequency distribution.

# Descriptive Statistics

## 8.1. Populations and samples

In statistics, the population is the entire group from which data may be collected and conclusions drawn. However, since populations may be very large and inconvenient to work with, statistical analysis is frequently performed on a sample, a smaller group drawn from the population. Assuming the sample is representative of the population, e.g. selected at random and sufficiently large, conclusions made about the sample should be valid for the population as a whole. For example, if we wanted to know whether children born in China in the 1950s were shorter than children born in China in the 1970s, it would be impossible to study the populations, i.e. all children born in China in the 1950s and 1970s, since these are far too large. However, statistical conclusions drawn about samples taken from these two populations should be valid for the whole population, assuming that the samples are unbiased and truly representative of the populations. So far so good. However, depending on the sample size, a statistic calculated for a sample based on the formula for a population may tend to produce a biased result, that is, an overestimate or an underestimate of the true value. For this reason, formulae for calculation statistics from samples often contain a small correction (e.g. $n - 1$ in place of $n$, the number of datapoints) to provide a more accurate answer. For this reason, you always need to be clear whether you are calculating a statistic for a population or a sample, and to use the correct formula (if appropriate).

## 8.2. The central tendency

Different frequency distributions can be described mathematically by measuring the central tendency and variability of the dataset. The central tendency is a summary measure of the middle of a dataset, which is commonly measured by any of three common descriptive statistics, the '3 Ms': mode, median and mean.

### Mode

The mode is the most frequently occurring value in a dataset. It is easy to determine, but is subject to great variation and consequently is of limited value.

### Median

The median is the middle value in a dataset, i.e. half the variables have values greater than the median and the other half values which are less. The median is less sensitive to outliers (extreme scores) than the mean and is thus a better measure than the mean for highly skewed distributions, such as family income. Note that the median equals the 50th percentile ($P_{50}$), i.e. the second quartile ($Q_2$).

### Mean

The mean is the average value of a dataset, i.e. the sum of all the data divided by the number of variables. The arithmetic mean is commonly called the 'average'. When the word 'mean' is used without a modifier, it usually refers to the arithmetic mean. The mean is a good measure of central tendency for symmetrical (e.g. normal) distributions, but can be misleading in skewed distributions since it is influenced by outliers. Therefore, other statistics such as the median may be more informative for distributions which are highly skewed. The mean, median and mode are equal in symmetrical frequency distributions. The mean is higher than the median in positively (right) skewed distributions and lower than the median in negatively (left) skewed distributions.

The formula for the arithmetic mean is:

$$\text{mean} = \frac{\sum X}{N}$$

where $\sum$ means 'sum'; $X$ are the raw datapoints; and $N$ is the number of scores (datapoints).

The geometric mean is the $n$th root of the product of the scores, for example, the geometric mean of the scores 1, 2, 3 and 4 is the 4th root of $1*2*3*4$, which is the 4th root of $24 = 2.21$. The geometric mean is less affected by extreme values than the arithmetic mean and is useful for some positively skewed distributions. However, the arithmetic mean is far more commonly encountered than the geometric mean.

## 8.3. Variability

Measurements in biology are frequently quite variable. There are many different sources for this variation, such as biological differences between individuals, resolution of measurement techniques and simple experimental error. It is important to be able to measure and describe the variability in datasets. While the central tendency is a summary measure of the middle of a dataset, variability (or dispersion) measures the amount of scatter in the dataset (e.g. Figure 8.1). Variability is commonly measured by three criteria: range, variance and standard deviation.

### Range

Range is the difference between the largest and the smallest value in the dataset. Although it is a crude measure of variability, it is easy to calculate and useful as an outline description of a dataset, for example in box and whisker plots (Section 8.8). However, since the range only takes into account two values from the entire dataset, it may be heavily influenced

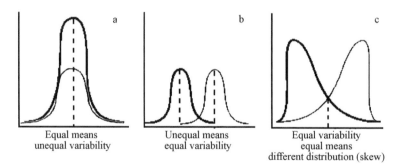

Figure 8.1   Variability

by outliers in the data. Therefore, another criterion is commonly used – the interquartile range, the interval between the 25th and 75th percentiles. In a normally distributed population, the interquartile range contains 50% of the datapoints making up the dataset: $Q_3 - Q_1$. A further measure which is even less subject to extreme scores is the semi-interquartile range, which is half of the interquartile range: $(Q_3 - Q_1)/2$.

Since the semi-interquartile range is little affected by extreme scores, it is a good measure of spread-out or skewed distributions. However, it is more subject to sampling fluctuation (i.e. how much a statistic varies from one sample to another) in normal distributions than the standard deviation (see below), so it not often used for data which are normally distributed.

## Variance

The variance of a dataset is more complicated to understand than the range, but is a measure of how 'spread-out' a distribution is. A deviation score is a measure of by how much each point in a frequency distribution lies above or below the mean for the entire dataset:

$$\text{Deviation score} = X - m$$

where $X$ is the raw score and $m$ is the mean for the dataset.

The variance is the mean of the squares of all the deviation scores for a dataset. This represents the amount of deviation of the entire dataset from the mean:

$$\text{Variance of a population} = \sigma^2 = \frac{\sum(X - \mu_x)^2}{N}$$

where $\sum$ is the sum, $X$ is the raw score, $\mu_x$ is the mean of the population, and $N$ is the number of datapoints.

$$\text{Variance of a sample} = s^2 = \frac{\sum(X-m)^2}{n-1}$$

where $\sum$ is the sum, $X$ is the raw score, $m$ is the mean of the sample, and $n$ is the number of datapoints in the sample.

Note that the variance is expressed in squared units, for example, if the raw scores are weight in kg, the variance is $kg^2$. For this reason, it is more useful to consider the square root of the variance, which is the standard deviation.

## Standard deviation

The standard deviation (SD) is the square root of the variance and is the most commonly used measure of how 'spread-out' a distribution is:

$$\text{Standard deviation of a population: } \sigma_x = \sqrt{\frac{\sum(X-\mu_x)^2}{N}}$$

$$\text{Standard deviation of a sample: } S_x = \sqrt{\frac{\sum(X-m)^2}{n-1}}$$

As with the other measures of data variability, the standard deviation determined from a sample (subset) of a dataset will be biased – since outliers are excluded, it will tend to underestimate the population standard deviation. Hence the formula needs to be modified for samples rather than whole populations. The standard deviation is probably the most useful measure of data spread. As you will see, many formulas in inferential statistics (Chapters 10 and 11) use the standard deviation. Although the standard deviation is less sensitive to extreme scores than the range, it is more sensitive than the semi-interquartile range. For this reason, the standard deviation should at least be supplemented by if not replaced by the semi-interquartile range when the possibility of extreme scores is present or for highly skewed datasets.

## 8.4. Standard error

Any statistic can have a standard error, which is the standard deviation of the sampling distribution of that statistic. Inferential statistics and significance testing (Chapters 10 and 11), and confidence intervals (below) are all based on standard errors. The standard deviation is an index of how closely individual data points cluster around the mean, thus each standard deviation refers to an individual datapoint. In contrast, standard errors indicate how much sampling fluctuation a summary statistic shows, that is, how good an estimate of the population the statistic is (e.g. the standard error of the mean, $\sigma_m$). The standard error of any statistic depends in part on the sample size – in general, the larger the sample size the smaller the standard error.

$$SE = \frac{SD}{\sqrt{N}}$$

How good an estimate is the mean of a population? One way to determine this is to repeat an experiment many times and to determine the mean of the means. However, this is at best tedious and frequently impossible. Fortunately, the standard error of the mean can be calculated from a single experiment and indicate the variability of the statistic:

$$\sigma_m = \frac{SD}{\sqrt{N}}$$

We will come back to the use of standard errors again later.

## 8.5. Confidence intervals

In a normal distribution 68% of datapoints fall within $\pm 1$ standard deviations from the mean; 95% of datapoints fall within $\pm 2$ standard deviations from the mean (actually $\pm 1.96$ standard deviations); and 99.7% of datapoints fall within $\pm 3$ standard deviations from the mean (Figure 8.2).

A confidence interval gives an estimated range of values which is likely to include an unknown datapoint. The width of the confidence interval gives us some idea about how uncertain we are about the parameter, for example, a very wide interval may indicate that more data should be collected before anything very definite can be said about the parameter. Confidence intervals are more informative than the results of inferential tests (Chapters 10 and 11), which only help you decide whether to reject a

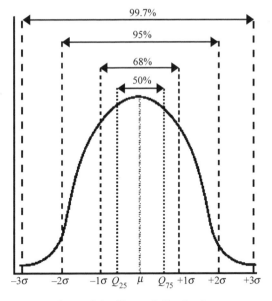

**Figure 8.2** Normal distribution

hypothesis, since they provide a probable range of numerical values for statistical parameters.

In a normal distribution, since there is less than a 1 in 20 chance of any datapoint falling outside ±2 standard deviations from the mean, we say that this range represents the 95% confidence interval, and the probability ($P$) of this range containing a particular datapoint is $P = 0.95$ (Chapter 9 contains a more detailed explanation of probability). Similarly, since there is less than a 1 in 99.7 chance of any sample in the population falling outside ±3 standard deviations; this represents a 99% confidence interval for the population, and $P = 0.99$. Confidence intervals can be constructed for any statistical parameter, not just the mean.

So when do you use standard deviations, standard errors or confidence intervals?

1. Use standard deviations when you are referring to individual data points. This tells you about the spread of the data.

2. Use standard errors when you are referring to differences between sample statistics, e.g. the mean. This tells you about the accuracy of your estimate.

3. Use confidence intervals when you want to convey the significance of differences between groups.

## 8.6. Parametric and non-parametric statistics

Statistical methods which depend on estimates of the parameters of populations or probability distributions are referred to as parametric methods, and include: Student's $t$-test; ANOVA (analysis of variance); regression analysis; and correlation analysis. These tests are only meaningful for continuous data which is sampled from a population with an underlying normal distribution, or whose distribution can be rendered normal by mathematical transformation. Non-parametric methods require fewer assumptions about a population or probability distribution and are applicable in a wider range of situations. For example, they can be used with qualitative data, and with quantitative data when no assumption can be made about the population probability distribution.

Non-parametric methods are useful in situations where the assumptions required by parametric methods are questionable. A few of the more commonly used non-parametric methods include: $\chi^2$ test; Wilcoxon signed-rank test; Mann–Whitney–Wilcoxon test; and Spearman rank correlation coefficient. These tests are 'distribution free', i.e. the population from which the sample was drawn does not need to have a normal distribution. Unlike parametric tests which can give erroneous results if applied to the 'wrong sort of data', these methods can be safely used in a wider range of circumstances. Unfortunately, they are less flexible in practice and less powerful than parametric tests.

> In cases where both parametric and non-parametric methods are applicable, statisticians usually recommend using parametric methods because they tend to provide better precision.

In statistical jargon, accuracy is a measurement of how close the average of a set of measurements is to the true or target value. Precision is a measure of the closeness of repeated observations to each other without reference to the true or target value, i.e. the reproducibility of the result.

# 8.7. Choosing an appropriate statistical test

In order to choose an appropriate statistical test, you must answer two questions:

1. What are the features of the dataset being analysed?

2. What is the goal of the analysis?

Table 8.1 summarizes some of the statistical tests which can be used to analyse different datasets. Not all of these tests are described in this book,

Table 8.1   Some of the statistical tests which can be used to analyse different datasets

| Goal | Dataset | | |
|---|---|---|---|
| | Normal distribution | Non-normal distribution | Binomial distribution |
| Describe one group | Mean, standard deviation | Median, interquartile range | Proportion |
| Compare one group to a hypothetical value | One-sample $t$-test | Wilcoxon test | $\chi^2$ or binomial test |
| Compare two unpaired groups | Unpaired $t$-test | Mann–Whitney test | Fisher's exact test (or $\chi^2$ for large samples) |
| Compare two paired groups | Paired $t$-test | Wilcoxon test | McNemar's test |
| Compare three or more unmatched groups | One-way ANOVA | Kruskal–Wallis test | $\chi^2$ test |
| Compare three or more matched groups | Repeated-measures ANOVA | Friedman test | Cochrane $Q$ test |
| Quantify association between two variables | Pearson correlation | Spearman correlation | Contingency coefficients |
| Predict value from another measured variable | Simple regression | Non-parametric regression | Simple logistic regression |
| Predict value from several measured variables | Multiple regression | | Multiple logistic regression |

but they have been included in the table for reference purposes. In subsequent chapters we will explore the most frequently employed statistical methods and how they can be used.

## 8.8. Exploratory data analysis

Hopefully, it will now be clear that one of the most important aspect of statistics is to use the appropriate method rather than a test which may generate a meaningless and misleading answer. Choosing a test largely depends on the nature of the data being analysed, and critically whether this has a normal distribution (so a parametric test can be used) or not (meaning a non-parametric method must be used). This crucial information can be obtained through a process known as *exploratory data analysis*, which includes many tools designed to reveal possible errors in the data (calculation or experimental errors, typing mistakes, etc.), data outliers, which should be investigated, and the underlying nature of the dataset (e.g. frequency distribution).

Exploratory data analysis comprises many different methods, including descriptive statistics, but the most powerful are graphical methods which literally paint a picture of the dataset. As an example, we will look at some of the most frequently used methods, all of which are easily performed by hand or with commonly available software.

### Scatter plots

Consider the three datasets in Table 8.2. At first sight, all three look very similar, with identical means and standard deviations. However, a scatter plot of the data quickly reveals considerable differences between the three datasets (see Figures 8.3 – 8.5).

### Frequency distribution histograms

A frequency distribution is a series of rectangles representing the frequencies of the class intervals. Since this was described in the previous chapter (Section 7.5), we will not repeat the description here.

Table 8.2 Datasets 1–3

| | Set 1 | | Set 2 | | Set 3 | |
| --- | --- | --- | --- | --- | --- | --- |
| | x1 | y1 | x2 | y2 | x3 | y3 |
| | 10 | 9.19 | 10 | 7.56 | 8 | 6.58 |
| | 8 | 8.14 | 8 | 6.67 | 8 | 6.66 |
| | 13 | 8.74 | 13 | 12.74 | 8 | 7.71 |
| | 9 | 8.81 | 9 | 7.11 | 8 | 8.84 |
| | 11 | 9.26 | 11 | 6.91 | 8 | 8.47 |
| | 14 | 8.10 | 14 | 8.84 | 8 | 7.04 |
| | 6 | 6.23 | 6 | 6.17 | 8 | 5.25 |
| | 4 | 3.10 | 4 | 6.39 | 19 | 12.56 |
| | 12 | 9.13 | 12 | 8.15 | 8 | 5.56 |
| | 7 | 7.26 | 7 | 6.42 | 8 | 7.91 |
| | 5 | 4.74 | 5 | 5.73 | 8 | 6.89 |
| Mean | 9.0 | 7.5 | 9.0 | 7.5 | 9.0 | 7.5 |
| Standard deviation | 3.3 | 2.0 | 3.3 | 2.0 | 3.3 | 2.0 |

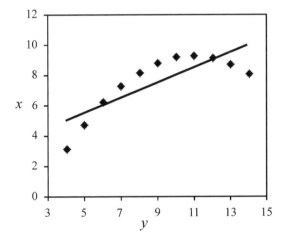

Figure 8.3 Dataset 1: the datapoints all lie on a smooth curve with little scatter – this would appear to be 'good' data

## Stem and leaf plots

A stem and leaf plot is like a histogram turned on its side but shows more information – the numerical values of each datapoint in addition to the

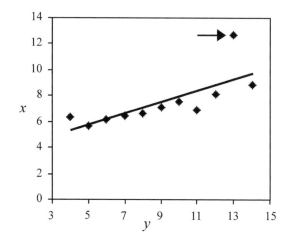

**Figure 8.4** Dataset 2: most of the datapoints lie close to a straight line, but one point (arrow) is suspiciously misplaced. This could be the result of either experimental or typographical error, but it is certainly worth investigating the cause before performing further analysis

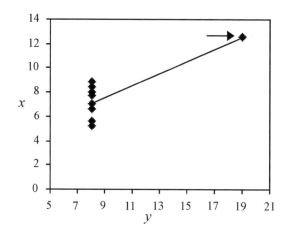

**Figure 8.5** Dataset 3: in this example, the single data outlier (arrow) would heavily influence the result of any statistical analysis. It is important to investigate the cause of this isolated datapoint (e.g. experimental error or design) and to consider carefully whether to include this datapoint in any analysis

overall pattern. The dataset – 39, 42, 44, 47, 48, 48, 51, 52, 53, 53, 54, 55, 55, 55, 55, 56, 56, 57, 57, 58, 58, 59, 59, 59, 59, 61, 61, 62, 63, 63, 64, 65, 65, 65, 66, 66, 66, 67, 69, 69, 71, 71, 76, 81, 84, 92 would be represented as:

```
3 | 9
4 | 2 4 7 8 8
5 | 1 2 3 3 3 4 5 5 5 5
5 | 6 6 7 7 8 8 8 9 9 9 9
6 | 1 1 2 3 3 4 5 5 5
6 | 6 6 6 7 9 9
7 | 1 1 6
8 | 1 4
9 | 2
```

The numbers to the left are the 'stem' of the plot – the tens digits in the frequency distribution of the dataset. The numbers to the right are the 'leaves' – the units digits in the frequency distribution, e.g. 5 | 6 represents a score of 56. Scores greater than 99 can be represented as follows: 25 | 6 for 256, 67 | 9 for 679, etc. The effect produced is that of a histogram, but each individual datapoint can be seen. No graphical software is necessary to produce the pattern, which can easily be reproduced in text form. In this example, the data approximates to a normal distribution (with a slight left-skew).

## Box and whisker plots

This alternative method of examining data makes use of common calculated numerical measures (median, interquartile range), but displays the data in a visual form (Figure 8.6). In the top plot in Figure 8.6, the median value is symmetrically placed in the middle of the box (interquartile range, which by definition covers 50% of the points in the dataset), so this dataset is normally distributed. In the middle plot, the interquartile range (box) is the same, but here the median value is at the right-hand end of the box, meaning that this dataset does not have a normal distribution, but has what is known as a right or positive skew. This dataset would provide inaccurate answers if subjected to parametric tests (unless transformed to a normal distribution first). In the lower plot, the median value is again in

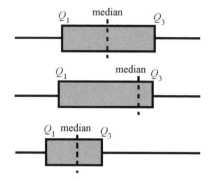

**Figure 8.6**   Box and whisker plots

the middle of the interquartile range, but this (i.e. the box) covers less of the dataset, meaning that there are more outliers in the data which could reduce the accuracy of any statistical analysis.

Examining the data in question using the above methods *before* applying any statistical tests is vital to any meaningful analysis, ensuring that any numerical summaries of the data or predictions made from the data are valid. Unfortunately, it is frequently overlooked. *Always* carry out some form of exploratory data analysis before proceeding further. Preferably draw at least one picture or graph. Much of statistics is about detecting patterns – something which the human eye and brain are very good at.

## Problems (answers in Appendix 1)

Table 8.3 contains a set of data on the microbiological quality of bottled drinking water. In this study, the number of bacterial colony-forming units per millilitre of bottled water was measured for 120 different water samples.

**8.1.** Construct a grouped frequency distribution table for this dataset.

**8.2.** Plot a frequency distribution histogram of the data.

**8.3.** How would you describe this dataset (normal, negative skew or positive skew?)

**8.4.** Calculate:

  (a) the 90th percentile for this dataset;

  (b) the 25th percentile for this dataset.

Table 8.3   Microbiological quality of drinking water

| Colony forming units $mL^{-1}$ | | | | | |
|------|------|------|------|------|------|
| 9159 | 6351 | 9726 | 8859 | 5832 | 6891 |
| 3783 | 7613 | 9527 | 9292 | 7512 | 6631 |
| 848  | 8799 | 8259 | 7645 | 9166 | 5864 |
| 7478 | 6758 | 6038 | 7952 | 8166 | 7078 |
| 7999 | 8492 | 8712 | 7718 | 8352 | 8659 |
| 8652 | 5791 | 8392 | 7698 | 8185 | 6951 |
| 8952 | 5184 | 8005 | 7912 | 4664 | 906  |
| 7818 | 9085 | 8292 | 7779 | 8259 | 8119 |
| 4117 | 6512 | 8432 | 8452 | 7545 | 383  |
| 8939 | 8672 | 6105 | 8966 | 8693 | 7532 |
| 9246 | 7598 | 6098 | 9413 | 8279 | 8252 |
| 4584 | 8686 | 7919 | 4504 | 6237 | 9146 |
| 6171 | 4184 | 8906 | 5097 | 7532 | 8586 |
| 6538 | 8793 | 6611 | 7879 | 6805 | 8246 |
| 7645 | 9092 | 8158 | 8339 | 8599 | 9006 |
| 7799 | 8659 | 7619 | 9166 | 8079 | 5084 |
| 2396 | 8365 | 8566 | 7478 | 8172 | 7812 |
| 5417 | 7685 | 8519 | 1735 | 8486 | 6905 |
| 8512 | 8079 | 7912 | 8653 | 7785 | 8699 |
| 6571 | 7732 | 8739 | 7798 | 7625 | 7519 |

**8.5.** Calculate:

  (a) the mean for this dataset;

  (b) the median for this dataset;

  (c) the mode for this dataset.

**8.6.** Calculate:

  (a) the range for this dataset;

  (b) the semi-interquartile range for this dataset;

(c) the variance for this dataset;

(d) the standard deviation for this dataset.

8.7. Exploratory data analysis:

(a) Construct a scatter plot of the following dataset. Are the data normally distributed?

| $x$ | 10 | 8 | 13 | 9 | 11 | 14 | 6 | 4 | 12 | 7 | 5 |
|---|---|---|---|---|---|---|---|---|---|---|---|
| $y$ | 11.2 | 8.1 | 9.7 | 9.8 | 12.8 | 8 | 6.2 | 3 | 11.8 | 7.3 | 5.7 |

(b) Construct a frequency histogram of the following dataset. Are the data normally distributed?

| $x$ | 1–10 | 11–20 | 21–30 | 31–40 | 41–50 | 51–60 | 61–70 | 71–80 | 81–90 | 91–100 |
|---|---|---|---|---|---|---|---|---|---|---|
| $y$ | 0 | 0 | 1 | 4 | 6 | 9 | 13 | 9 | 8 | 4 |

(c) Construct a stem and leaf diagram of the following dataset. Are the data normally distributed?

21, 23, 25, 26, 26, 27, 29, 1, 32, 32, 33, 35, 35, 36, 37, 38, 38, 39, 41, 41, 41, 42, 42, 44, 45, 47, 48, 48, 49, 51, 52, 53, 53, 53, 54, 55, 55, 55, 57, 61, 62, 63, 66, 71, 74, 91.

(d) Sketch a box and whisker plot of the following dataset. Are the data normally distributed? 14, 20, 22, 25, 27, 28, 31, 33, 38, 42, 51, 53, 61, 62, 65, 71, 74, 77, 78, 84, 86, 91. Median $= 52$, first quartile $= 29$, third quartile $= 73$.

# 9

# Probability

**LEARNING OBJECTIVES:**

On completing this chapter, you should understand the basic principles of probability theory, including:

- how to calculate probability in simple scenarios;
- the difference between selection with and without replacement;
- how to calculate the probability of multiple events.

## 9.1. Probability theory

Although most people find probability an interesting and enjoyable area of mathematics, why should a biologist need to understand and know how to calculate probabilities? This is because statistical methods depend upon probability theory.

Examples include important activities such as sampling from populations and hypothesis testing (Chapter 10), and probability distributions (Chapter 8).

$$\text{Probability, } P = \frac{\text{number of specific outcomes of a trial}}{\text{total number of possible outcomes of a trial}}$$

The simplest way to understand probabilities is through proportional frequency.

*Example*

In a group of mice there are 200 white mice and 50 brown mice:

1. Probability, $P$, is normally written as a decimal, e.g. $P = 0.5$. All probabilities lie between 0 and 1.

2. The proportional frequency of brown mice is $50/250 = 1/5 = 0.2$.

3. If we randomly take one mouse there is a $1/5$ chance of it being brown (0.2).

4. The probability of picking a brown mouse as a single random sample is equivalent to the proportional frequency of brown mice in the group (population).

5. If there were 250 white mice, the probability of selecting a brown mouse would be $0/250 = 0$. The probability of selecting a white mouse would be $250/250 = 1$.

## 9.2. Replacing or not replacing selections

If we replace the first selection from a population before making a second selection, then the probability of making any given selection is unaltered. Thus, in the above example the probability of picking a brown mouse is still $50/250 = 1/5 = 0.2$. However, if we do not replace our first selection the probability when making the second selection changes.

*Example*

In a group of mice there are 200 white mice and 50 brown mice:

1. Selection one = a brown mouse.

2. If this is not replaced there are now 249 mice (not 250) and only 49 brown mice (not 50). The probability of picking a brown mouse in the second sample is now $49/249$, not $50/250$. The chance of randomly selecting a brown mouse has decreased (slightly).

3. Similarly, the probability of randomly picking a white mouse in the second sample is now 200/249 rather than 200/250 as it would have been in the first selection, i.e. the chance of picking a white mouse in the subsequent selection increases as the chance of picking a brown mouse decreases.

Studying repeated samples (selections) from natural populations is easier if we assume that replacement occurs. This is usually true if the population is large, for example, taking one locust from a swarm of millions will not significantly change the overall population. When the result of the first sample does not affect the probability of the result of subsequent samples, the samples are said to be independent (an important requirement of many of statistical tests).

# 9.3. Calculating the probability of multiple events

There are two rules of probability:

1. **The SUM or OR rule** – the probability of any one of several distinct events is the sum of their individual probabilities, provided that the events are mutually exclusive (occurrence of one event precludes the others, e.g. selection without replacement).

2. **The PRODUCT or AND rule** – the probability of several distinct events occurring successively or jointly is the product of their individual probabilities, provided that the events are independent (the outcome of one event must have no influence on the others, e.g. tossing a coin).

The number of possible combinations of events is given by the factorial product of the number of events (written as '$n$') – the product of an integer and all the lower integers, for example, for three events ($X$, $Y$, $Z$), the number of possible combinations $= 3! = 3 * 2 * 1 = 6$:

| 1 | 2 | 3 | 4 | 5 | 6 |
|-----|-----|-----|-----|-----|-----|
| XYZ | XZY | YXZ | YZX | ZXY | ZYX |

Note that these are all different combinations, for example, crossing the road then looking for cars is not the same as looking for cars then crossing the road.

## Example

A population of 50 brown mice, 200 white mice, selections with replacement:

1. The probability of three brown mice in three selections:

$$(50/250) * (50/250) * (50/250)$$
$$= (1/5) * (1/5) * (1/5) = 0.008$$

2. The probability of selecting, in order, brown, brown and then white:

$$(50/250) * (50/250) * (200/250)$$
$$= (1/5) * (1/5) * (4/5) = 0.032$$

3. If, however, we are not interested in the order (i.e. brown, brown, white) but just the overall outcome (i.e. two brown, one white), the probability is different. The possible outcome of three selections with replacement is shown in Table 9.1. Thus, the sum of probabilities of a set of mutually exclusive, exhaustive outcomes is 1, but the probability of two brown mice and one white mouse, irrespective of the order of selection is as shown in Table 9.2. Note the difference in outcome between an ordered selection (probability $= 0.032$) and selection irrespective of order (probability $= 0.096$) $=$ the sum of all the possible ordered selections.

**Table 9.1**   Possible outcome of three selections with replacement

| Selection outcome | | | Probability of selection | | | Probability of outcome | |
|---|---|---|---|---|---|---|---|
| 1 | 2 | 3 | 1 | 2 | 3 | Sum | Total |
| B | B | B | 1/5 | 1/5 | 1/5 | $(1/5) * (1/5) * (1/5)$ | 0.008 |
| B | W | B | 1/5 | 4/5 | 1/5 | $(1/5) * (4/5) * (1/5)$ | 0.032 |
| B | B | W | 1/5 | 1/5 | 4/5 | $(1/5) * (1/5) * (4/5)$ | 0.032 |
| B | W | W | 1/5 | 4/5 | 4/5 | $(1/5) * (4/5) * (4/5)$ | 0.128 |
| W | B | B | 4/5 | 1/5 | 1/5 | $(4/5) * (1/5) * (1/5)$ | 0.032 |
| W | W | B | 4/5 | 4/5 | 1/5 | $(4/5) * (4/5) * (1/5)$ | 0.128 |
| W | B | W | 4/5 | 1/5 | 4/5 | $(4/5) * (1/5) * (4/5)$ | 0.128 |
| W | W | W | 4/5 | 4/5 | 4/5 | $(4/5) * (4/5) * (4/5)$ | 0.512 |
| | | | | | | Total | 1.0 |

**Table 9.2** Probability of two brown mice and one white mouse irrespective of order of selection

| Selection outcome | | | Probability of selection | | | Probability of outcome | |
|---|---|---|---|---|---|---|---|
| 1 | 2 | 3 | 1 | 2 | 3 | Sum | Total |
| B | W | B | 1/5 | 4/5 | 1/5 | $(1/5)*(4/5)*(1/5)$ | 0.032 |
| B | B | W | 1/5 | 1/5 | 4/5 | $(1/5)*(1/5)*(4/5)$ | 0.032 |
| W | B | B | 4/5 | 1/5 | 1/5 | $(4/5)*(1/5)*(1/5)$ | 0.032 |
| | | | | | | Total | 0.096 |

# 9.4. The binomial distribution

The binomial probability distribution describes what will happen when there are only two possible outcomes of an event, e.g. tossing a coin (heads or tails) or selections from a population consisting of two types of member (e.g. brown and white mice).

Such binary variables turn out to occur quite frequently in biology. In its simplest form, the binomial expansion summarizes the possible outcomes for any number of samples when there are only two possible outcomes (e.g. brown and white mice). For independent events, the binomial distribution is given by:

$$(P+Q)^n$$

where $P$ is the probability of one of the possible events, $Q$ is the probability of the second event ($Q = 1 - P$), and $n$ is the number of trials in the series.

For samples of 1 ($n = 1$): $(P+Q)^1 = (P+Q)$,

For samples of 2 ($n = 2$): $(P+Q)^2 = P^2 + 2PQ + Q^2$,

For samples of 3 ($n = 3$): $(P+Q)^3 = P^3 + 3P^2Q + 3PQ^2 + Q^3$, etc.

To return to the mice, these expansions of the binomial equation describe all the possible outcomes from the experiment above. If $P$ = brown mice and $Q$ = white mice, for three samples from the population ($n = 3$) there is: one way of obtaining three brown mice (BBB) $= P^3$; three ways of obtaining two brown mice and one white mouse (BBW:BWB:WBB) $= 3P^2Q$; three ways of obtaining one brown mouse and 2 white mice (BWW:WBW:WWB) $= 3PQ^2$; and one way of obtaining three white mice (WWW) $= Q^3$.

These are all the possible outcomes. In the population from which the samples were drawn:

$$50 \text{ brown mice, } P = 50/250 = 0.2$$
$$200 \text{ white mice, } Q = 200/250 = 0.8$$

and we can therefore calculate the distribution of outcomes from the binomial equation. In this example we can calculate the probability of two brown mice and one white mouse being selected as:

$$3P^2Q = 3(0.2)^2(0.8) = 0.096$$

This method is acceptable when there is a small number of samples and a small number of outcomes, but gets progressively more difficult as the sample size increases. For example, try using this method to calculate how many different ways there are to select seven brown mice and six white mice in 13 selections. To perform such calculations as this, we can use the following equation:

$$\text{Number of outcomes} = \frac{n!}{r!(n-r)!}$$

where $n$ is the number of selections and $r$ is the number of one of the outcomes (remember '!' = factorial).

## Example

For two brown mice and one white mouse (i.e. BBW, BWB, WBB), the number of outcomes is:

$$\frac{3!}{2!(3-2)!} = \frac{3*2*1}{2*1(1)} = \frac{6}{2} = 3$$

So for seven brown mice and six white mice, the number of possible outcomes is given by:

$$\frac{13!}{7!(13-7)!} = \frac{13!}{7!*6!} = \frac{13*12*11*10*9*8}{6*5*4*3*2*1}$$

$$= \frac{1\,235\,520}{720} = 1716$$

If we know the probability of the outcome for a single selection (e.g. probability of selecting a brown or a white mouse), we can calculate the total probability for the outcome using:

$$P(r) = \frac{n!}{r!(n-r)!} * p^r(1-p)^{n-r}$$

where $P$ is the total probability of the outcome (e.g. two brown mice and one white mouse), $p$ is the probability of the event that occurs $r$ times, and $(1-p)$ is the probability of the event that occurs $n-r$ times.

In our example of two brown mice and one white mouse:

$$\frac{3!}{2!(3-2)!} * \left(\frac{50}{250}\right)^2 * \left(\frac{200}{250}\right)^1$$
$$= 3 * (0.2)^2 * (0.8)^1 = 3 * 0.04 * 0.8 = 0.096$$

In practice, rather than actually performing such calculations, it is more usual to look up the probability of an event from a pre-calculated table of binomial probabilities (Appendix 3).

A particular importance of probability theory in statistics is that it controls sampling of populations and can be used to determine how large a sample needs to be taken from a population in order for an experiment to be successful, i.e. to have a statistically meaningful outcome.

## Example

Suppose that 4% of students carry an inherited defect in the (mythical) *statz* gene which restricts the ability of carriers to understand statistics. The only way to determine if someone is a carrier is to select individuals from the population at random and test them. If the number of students tested is too small there is a risk of not finding any carriers, but if it is too large, it will not be possible to mark all the tests. What sample size is required to give a good likelihood of sampling affected individuals? The binomial distribution can be used in a case such as this because the variable is binary, that is, each individual will or will not carry the defective gene. If 4% of students are carriers of the gene, then $P = 0.04$ (*statz*$^-$) and $Q = 0.96$ (*statz*$^+$). To find the probability of finding some (i.e. one or more) carriers of the gene, the most common method is to calculate is the probability of no cases [i.e. $P(0)$] for a given sample size, e.g. 10. Using the binomial equation, if the number of carriers, $r$, is 0, and the number of trials, $n$, is 10, we can calculate the probability of testing 10 individuals and finding no carriers:

$$P(r) = \frac{n!}{r!(n-r)!} * p^r(1-p)^{n-r}$$

$$P(0) = \frac{10!}{0!(10-0)!} * 0.04^0(1-0.04)^{10-0}$$

(NB. Any number raised to the power 0 is 1 and any number raised to the power 1 is itself, e.g. $20^0 = 1$ and $20^1 = 20$, so $1! = 0! = 1$.)

$$P(0) = 1 * 1 * 0.96^{10}$$
$$P(0) = 0.67$$

Thus, if 4% of students are carriers, there is a 67% chance that a sample of 10 students will fail to find any carriers. This tells us a sample size of 10 is too small to give a reasonable chance of finding at least one carrier, so we need to test a larger sample of the population:

1. If the number of students tested is 20, $P(0) = 0.96^{20} = 0.44$, i.e. there is now a 56% chance of finding an affected carrier ($1 - 0.44 = 0.56$).

2. If the number of students tested is 40, $P(0) = 0.96^{40} = 0.20$, i.e. a 80% chance of finding an affected carrier ($1 - 0.2 = 0.8$), etc.

Of course, the lower the frequency of any characteristic in a population, the higher the probability of not finding any positive results in a small sample. For example, if only 1% of students are $statz^+$ there is only a 10% chance of finding a carrier in a sample of size 10, i.e. $P(0) = 0.99^{10} = 0.9$. This method is useful to determine the minimum sample number needed to obtain at least one positive result from a sample for any binary variable, for example to find at least one affected carrier in a random sample. For other types of variable which may be continuous and normally distributed, the usual method of determining sample sizes to use the standard deviation (Chapter 8).

## 9.5. Coincidences

When working with large numbers (populations), probability theory has some unexpected results. Many apparently unexpected coincidences are merely the result of probability theory operating on very large populations, for example:

1. The chances of winning the UK National Lottery jackpot with a single ticket are about 14 million to one. What are the chances of someone who buys one Lotto ticket every week winning the jackpot twice within a year? Astronomical? Not necessarily. The chance of any single person (e.g. you) winning the jackpot twice within a year are approximately $10^{14}$ to one, but if 25 million people each buy one ticket every week, the chance of *anyone* winning the jackpot twice within a year are much greater – less than 100 to one.

2. What are the chances that someone else in a group of people has the same birthday as you?

$$P = 1 - (364/365 * 363/365 * 362/365 ...)$$

In a group of 22 people, there is a 50% chance that two people have the same birthday ($P = 0.5$). In a group of 120 people it is likely that someone else has the same birthday as you (work it out yourself).

3. In a large grassy field, the chances of putting your finger on a particular blade of grass are millions to one, but if you reach down and touch the ground, the chance of touching any blade of grass is nearly 100% ($P = 1$).

Why do 'coincidences' matter? They matter because, when you are trying to determine if an event is statistically significant or not, the seemingly logical 'expected' answer can be very misleading – events which might seem very unlikely to occur by chance can do precisely that if enough cases are involved. Consider a statistical analysis of whether banging your head against a hard surface can cure the common cold. Many studies of this problem are conducted, each with 95% confidence limit ($P = 0.05$). As soon as 20 studies have been performed, there will be, on average, at least one scientific paper published which proves that banging your head against a wall cures colds. Yet if you had a cold, what would you do?

## Problems (answers in Appendix 1)

**9.1.** Cystic fibrosis (CF) is the most common recessive genetic disorder in Caucasians – approximately one person in 2500 carries one copy of the CF gene, which occurs with equal frequency in males and females. If a couple are both carriers of the CF gene and have a child, the following probabilities apply: normal child, non-carrier, $P = 0.25$; normal child, carrier, $P = 0.50$; child with cystic fibrosis, $P = 0.25$. What is the probability that the couple will have:

(a) Two children (either sex) who do not carry the CF gene?

(b) One son who is a carrier?

(c) Two daughters, one who is a carrier and one who has cystic fibrosis?

(d) Two daughters with cystic fibrosis?

9.2. In order to study great crested newt (*Triturus cristatus*) populations, 150 newts are harmlessly marked with a temporary non-toxic dye. Fifteen newts are then returned to each of 10 ponds known to contain this species. One week later, the ponds are fished again and, of 351 newts caught, 54 are marked.

(a) Estimate the total population of great crested newts in these 10 ponds.

(b) If one pond has a population of 107 newts (15 marked), what is the probability of catching marked (M) and unmarked (U) newts in this order: UUMUUUMU?

9.3. In a health survey, 19 of 60 men and 12 of 40 women are found to smoke cigarettes.

(a) What is the probability of a randomly selected individual being a male who smokes?

(b) What is the probability of a randomly selected individual smoking?

(c) What is the probability of a randomly selected male smoking?

(d) What is the probability that a randomly selected smoker is male?

9.4. The probability of being infected with HIV from each single exposure to one of the following events is approximately: unprotected sexual intercourse with an HIV carrier, 0.005; sharing an infected needle for intravenous drug use, 0.007; needlestick injuries in healthcare workers, 0.003. The cumulative probability of being infected $P(i)$ after $n$ occurrences is given by the formula:

$$P(i) = 1 - (1 - k)^n$$

where $k$ is the probability of being infected with HIV from each single exposure and $n =$ the number of occurrences. What is the probability of being infected with HIV after:

(a) five occurrences of unprotected sexual intercourse with an HIV carrier;

(b) nine occurrences of sharing an infected needle for intravenous drug use;

(c) one needlestick injury in a healthcare worker who subsequently has unprotected sexual intercourse with an HIV carrier three times.

9.5. In a practical class, you are given three tubes of an enzyme (A B C) needed to perform an experiment you only have time to do once. A kind demonstrator has told you that only one of the tubes contains active enzyme – the other two are inactive. You choose tube A. To help you further, the demonstrator tells you that tube B contains inactive enzyme. Should you stick with tube A or switch to tube C for the experiment? Explain why.

# Inferential Statistics

**LEARNING OBJECTIVES:**

On completing this chapter, you should understand:

- how to draw reliable conclusions about samples taken from larger populations;
- how to compare different populations;
- when to use various inferential statistical methods;
- when *not* to use particular inferential statistical methods.

## 10.1. Statistical inference

To infer means to conclude from evidence. Statistical inference allows the formation of conclusions about almost any parameter of a sample taken from a larger population, for example, are conclusions based on a sample valid for the whole population? It also allows the formation of conclusions about the difference between populations with regard to any given parameter. There are two methods of reaching these sorts of statistical inference, estimation and hypothesis testing.

### Estimation

In estimation, a sample from a population is studied and an inference is made about the population based on the sample. The key to estimation is the probability with which particular values will occur during sampling; this allows the inference about the population to be made. The values which occur are inevitably based on the sampling distribution of the population. The key to making an accurate inference about a population therefore depends on random sampling, i.e. where each possible sample of the same size has the same probability of being selected from the population. In real life, it is often surprisingly difficult to take truly random samples from a population. Shortcuts are frequently taken, e.g. every third item on a list, 'expert' opinion, or simply taking the first $n$ results obtained. Estimation is a relatively crude method of making population inferences. A much better method and the one which is normally used in statistical analysis is hypothesis testing.

## Hypothesis testing

To answer a statistical question, the question is translated into a hypothesis – a statement which can be subjected to test. Depending on the result of the test, the hypothesis is accepted or rejected. The hypothesis tested is known as the null hypothesis ($H_0$). This must be in the form of a true/false statement. For every null hypothesis, there is an alternative hypothesis ($H_A$). Constructing and testing hypotheses is an important skill, but the best way to construct a hypothesis is not necessarily obvious:

1. If one of the two hypotheses is 'simpler' it is given priority so that a more 'complicated' theory is not adopted unless there is sufficient evidence against the simpler one (Occam's Razor: 'If there are two possible explanations always accept the simplest').

2. In general, it is 'simpler' to propose that there is no difference between two sets of results than to say that there is a difference.

3. The null hypothesis has priority and is not rejected unless there is strong statistical evidence against it.

The outcome of hypothesis testing is to 'reject $H_0$' or 'do not reject $H_0$'. If we conclude 'do not reject $H_0$', this does not necessarily mean that the null hypothesis is true, only that there is insufficient evidence against $H_0$ and in favour of $H_A$. Hypothesis testing never proves that the null hypothesis is true, just as rejecting the null hypothesis suggests but does not prove that the alternative hypothesis may be true.

In order to decide whether to accept or reject the null hypothesis, the level of significance ($\alpha$) required of the result must be decided. In general terms:

$\alpha = 0.05$ – significant (confidence interval 95%, $P = 1 - 0.95 = 0.05$), most commonly used;

$\alpha = 0.01$ – highly significant (confidence interval 99%, $P = 1 - 0.99 = 0.01$), strong statistical evidence;

$\alpha = 0.001$ – very highly significant (confidence interval 99.9%, $P = 1 - 0.999 = 0.001$), rarely used.

The level of significance allows us to state whether or not there is a 'significant difference' (note that this is a technical term which should only be used in the correct context) between populations, that is, whether any difference between populations is a matter of chance, due to experimental error, or so small as to be unimportant.

## 10.2. Procedure for hypothesis testing

1. Define $H_0$ and $H_A$, based on the guidelines given above.

2. Choose a value for $\alpha$. Note that this should be done before performing the test, not when looking at the result.

3. Calculate the value of the test statistic.

4. Compare the calculated value with a table of the critical values of the test statistic.

5. If the calculated value of the test statistic is *less than* the critical value from the table, accept the null hypothesis ($H_0$). Note that this does not mean that the null hypothesis has been conclusively proved, only that it has not been rejected.

6. If the calculated value of the test statistic is *greater than or equal to* the critical value from the table, reject the null hypothesis ($H_0$) and accept the alternative hypothesis ($H_A$).

Note that very small $P$-values (e.g. 0.001) do not signify large statistical differences, only that the observed differences are highly improbable given the null hypothesis tested. $P$-values indicate how sure you can be that there is a real difference, not the size of the difference. For example, a very small $P$-value can arise when any difference is tiny but the sample sizes very large. Conversely, a large $P$-value can arise when the effect is large but the sample size is small.

## 10.3. Standard scores ($z$-scores)

$z$-scores define the position of a score in relation to the mean using the standard deviation as a unit of measurement. They are therefore useful for comparing datapoints in different distributions.

$$z = (\text{score} - \text{mean})/\text{standard deviation}$$

The $z$-score is the number of standard deviations by which the score departs from the sample mean. Since this technique normalizes distributions, $z$-scores can be used to compare data from different sets, e.g. a student's performance on two different exams (e.g. did Joe Blogg's performance on test 1 and test 2 improve or decline?):

1. Joe B scored 71.2% on exam 1 (mean $= 65.4\%$, SD $= 3.55$) $z = (71.2 - 65.4)/3.55 = 1.63$.

2. Joe B scored 66.8% on exam 2 (mean $= 61.1\%$, SD $= 2.54$) $z = (66.8 - 61.1)/2.54 = 2.24$.

3. Conclusion – Joe B did better, compared with the rest of his classmates, on exam 2 than on exam 1, even though his mark was lower in the second exam.

Note that the $z$-score is a parametric statistic (Chapter 8), and is only meaningful when it refers to a normal distribution – calculating a $z$-score from a skewed dataset may not produce a meaningful number. Comparing $z$-scores for different distributions is also meaningless unless: the datasets being compared are as similar as possible (e.g. response to different doses of a drug under the same physiological conditions); and the shapes of the distributions being compared are as similar as possible.

## 10.4. Student's *t*-test (*t*-test)

Biological systems are complex, with many different interacting factors. To compensate for this, the most common experimental design in biology involves comparing experimental results with those obtained under control conditions. To interpret this type of experiment, we must be able to make objective decisions about the nature of any differences between the experimental and control results – is there a statistically significant difference or are the results due to experimental error or random chance (e.g. sampling error)? A frequently used test of statistical significance is Student's *t*-test (or simply *t*-test), devised by William Gosset ('Student') in 1908. The *t*-test is used to compare two groups and has two variants:

1. Paired *t*-test – used when each data point in one group corresponds to a matching data point in the other group.

2. Unpaired *t*-test – used whether or not the groups contain matching datapoints.

The *t*-test is a parametric test which assumes that the data analysed:

- Is continuous, interval data comprising a whole population or is sampled randomly from a larger population.

- Has a normal distribution (Chapter 8).

- If the sample size ($n$) is $< 30$, the variances (Chapter 8) of the two groups should be similar (*t*-tests can be used to compare groups with different variances if $n > 30$).

- The sample size should not differ hugely between the groups (e.g. $< 50\%$).

If you use the *t*-test under other circumstances, the results may be misleading. In other situations, non-parametric tests should be used to compare the groups, for example, the Wilcoxon signed rank test for paired data and the Wilcoxon rank sum test or Mann–Whitney test for unpaired data (not covered in this book). The *t*-test can only be used to compare two groups. To compare three or more groups, other tests must be used, for example, analysis of variance between groups (ANOVA; see Section

10.5). In general though, the *t*-test is quite robust and produces approximately correct results in many circumstances.

The paired *t*-test is used to investigate the relationship between two groups where there is a meaningful one-to-one correspondence between the data points in one group and those in the other, for example a variable measured at the same time points under experimental and control conditions. It is not sufficient that the two groups simply have the same number of datapoints. The advantage of the paired *t*-test is that the formula procedure involved is fairly simple.

## Procedure

1. Start with the hypothesis ($H_0$) that the mean of each group is equal, that is, there is no significant difference between the means of the two groups, e.g. control and experimental data. The alternative hypothesis ($H_A$) is therefore that the means of the groups are not equal. We test this by considering the variance (standard deviation) of each group.

2. Set a value for $\alpha$ (significance level, e.g. 0.05).

3. Calculate the difference for each pair (i.e. the variable measured at the same time point under experimental and controlled conditions).

4. Plot a histogram of the differences between data pairs to confirm that they are normally distributed – if not, stop.

5. Calculate the mean of all the differences between pairs ($d_{av}$) and the standard deviation of the differences (SD).

6. The value of *t* can then be calculated from the following formula:

$$t = \frac{d_{av}}{SD/\sqrt{N}}$$

where $d_{av}$ is the mean difference, i.e. the sum of the differences of all the datapoints (set 1 point 1 – set 2 point 1, etc.) divided by the number of pairs; SD is the standard deviation of the differences between all the pairs; and $N$ is the number of pairs. NB. The sign of *t* ($+/-$) does not matter; assume that *t* is positive.

7. The calculated value of *t* can then be looked up in a table of the *t* distribution (Appendix 3, or obtained from appropriate software). To do this, you need to know the 'degrees of freedom' (df) for the test. The result of any statistical test is influenced by the population size, for example it is more accurate to make 200 measurements than 20 measurements. Since the number of observations (population size) affects the value of statistics such as *t*, when we calculate or look up *t*, we need to take the population size into account – this is what degrees of freedom does. For a paired *t*-test:

$$df = n - 1 \text{ (number of pairs} - 1)$$

To look up *t*, you also need to determine whether you are performing a one-tailed or two-tailed test. In any statistical test we can never be 100% sure that we have to reject (or accept) the null hypothesis. There is therefore the possibility of making an error as shown in Table 10.1.

**Table 10.1** The possibility of making an error

| | | Null hypothesis | |
|---|---|---|---|
| | | True | False |
| Decision | Reject | Type I error | Correct |
| | Accept | Correct | Type II error |

Falsely rejecting a true null hypothesis is called a type I error. The probability of committing a type I error is always equal to the significance level of the test, $\alpha$. Failure to reject a false null hypothesis is called a type II error. The 'power' of a statistical test refers to the probability of correctly claiming a significant result. As scientists are generally cautious, it is considered 'worse' to make a type I error than a type II error; we thus reduce the possibility of making a type I error by having a stringent rejection limit, 5% ($\alpha = 0.05$). However, as we reduce the possibility of making one type of error, we increase the possibility of making the other type. Whether you use a one- or two-tailed test depends on your testing hypothesis.

1. One-tailed test – used where there is some basis (e.g. previous experimental observation) to predict the direction of the difference, such as expectation of a significant difference between the groups. In some circumstances, one-tailed tests can be valuable, for example if it is proposed that a new drug is more effective in the treatment of a disease than an existing drug. The new drug should only be adopted if there is a significant improvement in treatment outcome.

2. Two-tailed test – used where there is no basis to assume that there may be a significant difference between the groups; this is the test most frequently used. The result of a two-tailed test does not tell you if any difference between groups is 'greater than' or 'less than', only that there is a significant difference.

They are called 'tails' because of the region of retention and regions of rejection on a graph of the distribution of the test statistic (Figure 10.1). Note that the alternative hypothesis states 'there is a difference'; it does not state why there is a difference or whether the difference between the two groups is 'greater than' or 'less than'. If the alternative hypothesis had specified the nature of the difference, this would have been a one-tailed hypothesis. However, if the alternative hypothesis does not specify the nature of the difference, we can accept either a reduction or an increase and it is therefore a two-tailed hypothesis. For a variety of reasons two-tailed hypotheses are safer than one-tailed. Statistical tables are sometimes tabulated only for one-tailed hypotheses. To convert them to two-tailed, double the value of $\alpha$. A table of critical values of $t$ for Student's $t$ distribution is given in Appendix 3.

If the calculated value of $t$ is greater than or equal to the critical value of the test statistic, the null hypothesis is rejected, that is, there is evidence of a statistically significant difference between the groups. If the calculated value of $t$ is less than the critical value, the null hypothesis is accepted – there is no evidence of a statistically significant difference between the two groups.

The unpaired $t$-test does not require that the two groups be paired in any way, or even of equal sizes. A typical example might be comparing a variable in two experimental groups of patients, one treated with drug A

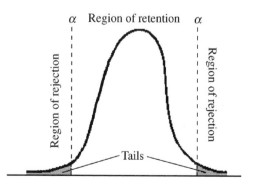

**Figure 10.1** 'Tails'

and one treated with drug B. Such situations are common in medicine where an accepted treatment already exists and it would not be ethical to withhold this from a control group. Here, we wish to know if the differences between the groups are 'real' (statistically significant) or could have arisen by chance. The calculations involved in an unpaired *t*-test are slightly more complicated than for the paired test. Note that the unpaired *t*-test is equivalent to one-way ANOVA (Section 10.5), used to test for a difference in means between two groups.

$$t = \frac{\bar{X}_A - \bar{X}_B}{\sqrt{(SE_A)^2 + (SE_B)^2}}$$

where $\bar{X}$ is the mean of groups A and B, respectively, and $SE = SD/\sqrt{N}$.
    For an unpaired *t*-test:

$$df = (nA + nB) - 2$$

where *n* is the number of values in the two groups being compared. Note that this is different from the calculation of the number of degrees of freedom for a paired *t*-test. Compare the calculated value of *t* with the critical value in a table of the *t* distribution (Appendix 3). Remember that the sign of *t* ( + / − ) does not matter, and assume that *t* is positive. If the calculated value of *t* is greater than or equal to the critical value, the null hypothesis is rejected – there is evidence of a statistically significant difference between the groups. If the calculated value of *t* is less than the critical value, the null hypothesis is accepted – there is no evidence of a statistically significant difference between the groups.

## Example

Consider the data from the following experiment. A total of 12 readings were taken, six under control and six under experimental conditions (Table 10.2). Before starting to do a *t*-test, several questions must be answered:

1. Are the datapoints for the control and experimental groups paired?

    No, they are just replicate observations, so we need to perform an unpaired *t*-test.

2. Are the data normally distributed?

**Table 10.2** Experimental data

|  | Experimental, group A | Control, group B |
|---|---|---|
|  | 11.2 | 10.3 |
|  | 13.1 | 12.6 |
|  | 9.3 | 8.4 |
|  | 10.2 | 9.3 |
|  | 9.6 | 10.8 |
|  | 9.8 | 8.9 |
| Variance | 1.68 | 1.96 |
| SD | 1.30 | 1.40 |
| $SE = SD/\sqrt{N}$ | 0.53 | 0.57 |

Yes, approximately:

|  | Experimental | Control |
|---|---|---|
| Mean | 10.53 | 10.05 |
| Median | 10.00 | 9.80 |

3. $H_0$: 'There is no difference between the populations of measurements from which samples have been drawn' ($H_A$: there is a difference).

4. Set the value of $\alpha = 0.05$ (i.e. a 95% confidence interval).

5. Are the variances of the two groups similar?

   Yes, approximately (1.68 vs 1.96).

6. Since all the requirements for a *t*-test have been met, we can proceed:

$$t = \frac{10.53 - 10.05}{\sqrt{(0.53)^2 + (0.57)^2}} = 0.62$$

7. Is this a one-tailed or a two-tailed test?

   Two-tailed, since we have no firm basis to assume the nature of any difference between the groups.

8. How many degrees of freedom are there?

$$df = (n_A - 1) + (n_B - 1) = 10$$

9. From the table of critical values of $t$ (Appendix 3), we can see that for a two-tailed test with $df = 10$ and $\alpha = 0.05$, $t$ would have to be 2.228 or greater for $> 5\%$ (0.05) of pairs of samples to differ by the observed amount.

10. Since $t_{calc} = 0.62$ and $t_{crit} = 2.228$, the null hypothesis is accepted. The conclusion is that there is no evidence of a statistically significant difference (at the 95% confidence level) between the experimental and the control groups in this experiment.

## 10.5. Analysis of variance (ANOVA)

Student's $t$-test can only be used for comparison of two groups. Although it is possible to perform many pair-wise comparisons to analyse all the possible combinations involving more than two groups, this is undesirable because it is tedious, but more importantly because it increases the possibility of type I errors (Section 10.4). However, ANOVA can compare two or more groups. ANOVA is a parametric test which assumes that:

1. The data analysed is continuous, interval data comprising a whole population or sampled randomly from a population.

2. The data has a normal distribution. Moderate departure from the normal distribution does not unduly disturb the outcome of ANOVA, especially as sample sizes increase, but highly skewed datasets result in inaccurate conclusions.

3. The groups are independent of each other.

4. The variances in the groups should be similar. For ANOVA, this is more important to accuracy that normal distribution of the data.

5. For two-way ANOVA, the sample size the groups is equal (for one-way ANOVA, sample sizes need not be equal, but should not differ hugely between the groups). This is because the results of ANOVA

tests can be upset by different variances in the groups, but this effect is minimized if the groups are of the same or similar sizes.

ANOVA tests come in various forms:

1. One-way (or one-factor) ANOVA – tests the hypothesis that means from two or more samples are equal (drawn from populations with the same mean). Student's $t$-test is actually a particular application of one-way ANOVA (two groups compared) and results in the same conclusions.

2. Two-way (or two-factor) ANOVA – simultaneously tests the hypothesis that the means of two variables ('factors') from two or more groups are equal (drawn from populations with the same mean), for example the difference between a control and an experimental variable, or whether there is a difference between alcohol consumption and liver disease in several different countries. It does not include more than one sampling per group. This test allows comments to be made about the interaction between factors as well as between groups.

3. Repeated measures ANOVA – used when members of a random sample are measured under different conditions. As the sample is exposed to each condition, the measurement of the dependent variable is repeated. Using standard ANOVA is not appropriate because it fails to take into account correlation between the repeated measures, violating the assumption of independence. This approach can be used for several reasons, such as where research requires repeated measures, for example, longitudinal research which measures each sample member at each of several ages – age is a repeated factor.

The $F$-ratio ('Fisher ratio') compares the variance within sample groups ('inherent variance') with the variance between groups ('treatment effect') and is the basis for ANOVA:

$$F = \text{variance between groups/variance within sample groups}$$

ANOVA works by comparing the relationship between the variability within groups, across groups and the total. The actual ANOVA calculation itself is quite laborious and best performed using statistical software (Appendix 2). If you insist on knowing the equations involved, they can be looked up in a statistics textbook or software manual. This chapter will concentrate on how to use ANOVA. The basic procedure is similar to that for performing a *t*-test:

1. Formulate the null hypothesis, i.e. that the means of the groups are equal.

2. Choose a confidence interval and set the significance level accordingly, e.g. $CI = 95\%$, $\alpha = 0.05$.

3. Calculate the test statistic ($F$) (best done using software).

4. Compare the calculated value of $F$ with a table of critical values of $F$.

5. If the calculated value of the $F$ is less than the critical value from the table, accept the null hypothesis ($H_0$). If the calculated value of $F$ is greater than or equal to the critical value from the table, reject the null hypothesis ($H_0$) and accept the alternative hypothesis ($H_A$).

## Examples

Tables 10.3–10.5 show an example of one-way ANOVA. The null hypothesis is that there is no difference between the four groups being

Table 10.3   Experimental data

| Pain Score for three analgesics | | | |
|---|---|---|---|
| Aspirin | Paracetemol (Acetaminophen) | Ibuprophen | Control (no drug) |
| 5 | 4 | 4 | 5 |
| 4 | 4 | 4 | 5 |
| 5 | 3 | 5 | 5 |
| 3 | 4 | 3 | 4 |
| 5 | 5 | 3 | 5 |
| 5 | 3 | 5 | 5 |
| 4 | 4 | 3 | 5 |

Table 10.4   Summary

| Groups | Count | Sum | Average | Variance |
|---|---|---|---|---|
| Aspirin | 7 | 31 | 4.43 | 0.62 |
| Paracetemol | 7 | 27 | 3.86 | 0.48 |
| Ibuprophen | 7 | 27 | 3.86 | 0.81 |
| Control (no drug) | 7 | 34 | 4.86 | 0.14 |

**Table 10.5  ANOVA**

| Source of variation | SS | df | F | $F_{crit}$ |
|---|---|---|---|---|
| Between groups | 4.96 | 3 | 3.23 | 3.01 |
| Within groups | 12.29 | 24 | | |
| Total | 17.25 | 27 | | |

compared. In this example, with a significance level of 95% ($\alpha = 0.05$), since the calculated value of $F$ (3.23) is greater than $F_{crit}$ (3.01), we reject the null hypothesis that the three drugs perform equally. The null hypothesis would have been rejected if even one of the groups differed significantly from the other three. A *post-hoc* comparison or series of individual pair-wise comparisons would have to be performed to determine which pair or pairs of means caused rejection of the null hypothesis, but since this was not part of the original question, we cannot address this directly here. If ANOVA is performed on three or more groups and it finds a significant difference, then a *post-hoc* test (also called pair-wise comparisons, multiple comparison tests, and multiple range tests) needs to be performed in order to make multiple comparisons between the groups. By comparing pairs of groups in every possible combination, the differences among them are revealed. There are various *post-hoc* tests which can be used, such as the 'Bonferonni', 'Scheffe', 'Tukey' and 'LSD' (least

**Table 10.6  Experimental results**

Apple codling moth (Cydia pomonella) caught in pheromone traps

| | Bait 1 | Bait 2 |
|---|---|---|
| Orchard 1 | 19 | 20 |
| | 22 | 22 |
| | 19 | 18 |
| | 18 | 19 |
| | 20 | 19 |
| | 21 | 20 |
| Orchard 2 | 22 | 21 |
| | 19 | 19 |
| | 19 | 18 |
| | 18 | 18 |
| | 20 | 20 |
| | 21 | 22 |

Table 10.7   Anova: two-factor without replication

| Summary | Count | Sum | Average | Variance |
|---|---|---|---|---|
| Orchard 1 | 2 | 39 | 19.5 | 0.5 |
|  | 2 | 44 | 22 | 0 |
|  | 2 | 37 | 18.5 | 0.5 |
|  | 2 | 37 | 18.5 | 0.5 |
|  | 2 | 39 | 19.5 | 0.5 |
|  | 2 | 41 | 20.5 | 0.5 |
| Orchard 2 | 2 | 43 | 21.5 | 0.5 |
|  | 2 | 38 | 19 | 0 |
|  | 2 | 37 | 18.5 | 0.5 |
|  | 2 | 36 | 18 | 0 |
|  | 2 | 40 | 20 | 0 |
|  | 2 | 43 | 21.5 | 0.5 |
| Bait 1 | 12 | 238 | 19.83 | 1.97 |
| Bait 2 | 12 | 236 | 19.67 | 2.06 |

Table 10.8   ANOVA

| Source of variation | SS | df | F | $F_{crit}$ |
|---|---|---|---|---|
| Rows | 40.5 | 11 | 10.57 | 2.82 |
| Columns | 0.17 | 1 | 0.48 | 4.84 |
| Error | 3.83 | 11 |  |  |
| Total | 44.5 | 23 |  |  |

significant difference) tests. It is beyond the scope of this chapter to go into *post-hoc* tests, so you will need to consult other sources or software manuals if you are ever in a position to need such tests.

Tables 10.6–10.8 show an example of two-way ANOVA. As always, the null hypothesis is that there is no difference between the groups being compared. In this example, with a significance level of 95% ($\alpha = 0.05$), the calculated value of $F$ (10.57) for the table rows (orchard 1 vs orchard 2) is greater than $F_{crit}$ (2.82), so the hypothesis that there is no difference between the orchards is rejected. However, the calculated value of $F$ (0.48) for the table columns (bait 1 vs bait 2) is less than $F_{crit}$ (4.84), so the hypothesis that there is no difference between the pheromone baits is accepted. This example only compares two groups, so it is relatively easy to interpret the outcome.

## 10.6. $\chi^2$-test

This is an example of a non-parametric test which, unlike Student's $t$-test and ANOVA, makes no assumptions about the distribution of the data. $\chi^2$ (pronounced 'kye-squared') is used when data consists of nominal or ordinal variables rather than quantitative variables, when we are interested in how many members fall into given descriptive categories (not for quantitative measurements, such as weight, etc.).

The $\chi^2$-test of independence asks 'Are two variables of interest independent (not related) or related (dependent)?' and deals with nominal and ordinal variable expressed as integers, that is, variables which fall into different, mutually exclusive categories. This is distinct from the $t$-test, which deals with interval variables, although the ANOVA test can also be performed on nominal data (Chapter 7). The $\chi^2$-test investigates whether the proportions of certain categories are different in different groups. When the variables are independent, knowledge of one variable gives no information about the other variable. When they are dependent, knowledge of one variable is predictive of the value of the other variable, for example:

1. Is level of education related to level of income?

2. Is membership of a political party related to a person's preferred television station?

3. Is there a relationship between gender and examination performance?

The $\chi^2$-test has two main uses: comparing the distribution of one category variable (nominal or ordinal) with another; and comparing an observed distribution with a theoretically expected one.

The expectation might be that the data would be normally distributed, or that particular attributes (e.g. treatment and disease) are independent, meaning there is no closer association than might be expected by chance. In the first case, a table of values for a normal distribution would be the source of the expected values. In the second, the expected values would be calculated assuming independence (random distribution). The $\chi^2$-test is a non-parametric test which assumes that the data analysed:

1. Consist of nominal or ordinal variables.

2. Consist of entire populations or are randomly sampled from the population.

3. No single data point should be zero (if so, use Fisher's exact test; Section 10.7).

4. All the objects counted should be independent of one another.

5. Eighty per cent of the expected frequencies should be 5 or more (if not, try aggregating groups or use Fisher's exact test for small sample sizes; Section 10.7).

If you use the $\chi^2$-test under other circumstances, the results may be misleading. The $\chi^2$-test is by default one-tailed and can only be carried out on raw data (not percentages, proportions or other derived data). The basis of the $\chi^2$-test is:

$$\chi^2 = \sum \frac{(\text{observed frequency} - \text{expected frequency})^2}{\text{expected frequency}}$$

Note that acceptance or rejection of the null hypothesis can only be interpreted strictly in terms of the question asked, for example 'There is a difference between the observed and expected frequencies' or 'There is no difference between the groups' and not extrapolated to 'There is a difference between the observed and expected frequencies because ...'.

## Example A: comparing the distribution of one category variable with another

Of 120 male and 100 female applicants to a university, 90 male and 40 female had work experience. Does the gender of an applicant to university correspond to whether or not they have prior work experience?

The starting point for most $\chi^2$ analyses of this type is to construct a contingency table, a table showing how the values of one variable are related to ('contingent on') the values of one or more other variables:

|  |  | Work experience | | |
| --- | --- | --- | --- | --- |
|  |  | Yes | No | Total |
| Gender of applicant | Male | 90 | 30 | 120 |
|  | Female | 40 | 60 | 100 |
|  | Total | 130 | 90 | 220 |

Next, formulate the null hypothesis ($H_0$): male and female applicants have equivalent work experience ($H_A$, male and female applicants have different work experience). Set a confidence interval, e.g. CI = 95%, so $\alpha = 0.05$. Calculate $x^2$:

$$x^2 = \sum \frac{(\text{observed frequency} - \text{expected frequency})^2}{\text{expected frequency}}$$

|  |  | Work experience | | |
| --- | --- | --- | --- | --- |
|  |  | Yes | No | Total |
| Gender of applicant | Male | $a$ | $b$ | $a+b$ |
|  | Female | $c$ | $d$ | $c+d$ |
|  | Total | $a+c$ | $b+d$ | $n$ |

$$x^2 = \frac{n(ad - bc)^2}{(a + b)(c + d)(a + c)(b + d)}$$

$$x^2 = \frac{220(90 \times 60 - 30 \times 40)^2}{(90 + 30)(40 + 60)(90 + 40)(30 + 60)}$$

$$x^2 = \frac{220(5400 - 1200)^2}{120 \times 100 \times 130 \times 90}$$

$$x^2 = \frac{3\,880\,800\,000}{140\,400\,000}$$

$$x^2 = 27.64$$

As explained earlier, the distribution of $x^2$ depends upon the number of degrees of freedom (df) in the test:

$$df = (\text{number of columns} - 1) * (\text{number of rows} - 1)$$

For the above test, $df = (2 - 1) * (2 - 1) = 1$. Look up the calculated value of $x^2$ in a table of critical values of the $x^2$ distribution (Appendix 3). If the calculated value of $x^2$ is greater than the critical value of $x^2$ (from the table), reject $H_0$. If the calculated value of $x^2$ is less than the critical value of $x^2$ (from the table), accept $H_0$.

In this example, $x^2 = 27.64$, greater than the critical value for 1 df, so the null hypothesis is rejected – male and female applicants do not have equivalent work experience. Note that, from the test result alone, we cannot say whether males or females have greater work experience, only that the two groups are not equal. In this example, it is fairly easy to work out which group has greater work experience by simply scrutinizing the table. The $\chi^2$-test has simply proved that the difference between the two groups is statistically significant (at a 95% confidence level). Of course, the differences between groups are not always as clear-cut as in this example.

## Alternative method: $\chi^2$ calculation using observed and expected values

An alternative method of calculating $\chi^2$ for the above example is to calculate the expected distributions assuming the null hypothesis to be true: 130 students out of a total of 220 had work experience. If the proportion of males and females with work experience were equivalent we would expect: males with experience $= (130/200) * 120 = 71$. Table 10.9 shows the contingency table.

Table 10.9   Contingency table

|  | Observed (O) | | Expected (E) | | O − E | | $(O − E)^2/E$ | |
|---|---|---|---|---|---|---|---|---|
|  | Yes | No | Yes | No | Yes | No | Yes | No |
| Male | 90 | 30 | 71 | 49 | 19 | − 19 | 5.1 | 7.4 |
| Female | 40 | 60 | 59 | 41 | − 19 | 19 | 6.1 | 8.8 |
| Total | 130 | 90 | 130 | 90 | 0 | 0 | 11.2 | 16.2 |

$\chi^2 = 11.2 + 16.2 = 27.4$. From the table of critical values of $\chi^2$, the calculated value is greater than the critical value, so the null hypothesis is rejected. The advantage of this method is that it can be applied to problems where there are more than two groups, for example:

1. Each of a group of 1350 students were immunized with one of five influenza vaccines under test. Is there any evidence that any one influenza vaccine is better than the others based on the numbers of students who developed influenza and those who did not?

2. We can produce a table with observed and expected values (not shown here). The overall $\chi^2$ value will inform us whether there are differences between the vaccines.

3. The sums of the $(O - E)^2/E$ for each vaccine will provide information about the contribution of each vaccine to the overall $\chi^2$ – the vaccine contributing the most to the overall difference will have the largest $(O - E)^2/E$.

## Example B: comparing an observed distribution with a theoretically expected one

Using the method of observed and expected values we can use the $\chi^2$-test to compare an observed distribution with a theoretically expected one. For example, in a population of mice:

| Colour | Observed | Expected from genetic theory |
|--------|----------|------------------------------|
| White  | 380      | 51%                          |
| Brown  | 330      | 40.8%                        |
| Black  | 74       | 8.2%                         |

Do the proportions observed differ from those expected? Formulate the null hypothesis ($H_0$): the observed distribution does not differ from the expected distribution ($H_A$, the observed distribution differs from the expected distribution). Set a confidence interval, e.g. CI $= 95\%$, so $\alpha = 0.05$. Table 10.10 shows the contingency table.

Table 10.10   Contingency table

| Colour | Observed | Theoretical proportion | Expected | $O - E$ | $(O - E)^2/E$ |
|--------|----------|------------------------|----------|---------|----------------|
| White  | 380      | 0.510                  | 400 (0.510 ∗ 784) | − 20 | 1.0 |
| Brown  | 330      | 0.408                  | 320 (0.408 ∗ 784) | 10 | 0.3125 |
| Black  | 74       | 0.082                  | 64 (0.082 ∗ 784) | 10 | 1.5625 |
| Total  | 784      | 1.0                    | 784      | 0       | 2.8750 |

Calculate $\chi^2 = 2.875$. Calculate df:

$$df = (\text{number of columns} - 1) * (\text{number of rows} - 1)$$

$$(\text{columns} = \text{observed, expected} = 2; \; \text{rows} = \text{white, brown, black} = 3)$$

$$= (2 - 1) * (3 - 1) = 1 * 2 = 2$$

From the table of critical values of $\chi^2$ (Appendix 3), the calculated value of $\chi^2$ is less than the critical value, so the null hypothesis is accepted.

Although the $\chi^2$-test is, strictly speaking, non-parametric, it still has limitations. All the objects counted should be independent of one another, so the outcome of counting one should not influence the outcome of counting any of the others. Eighty per cent of the expected frequencies should be 5 or more. If this is not the case, it is sometimes possible to get around this difficulty by aggregating (combining) groups. Also, no single data point should be zero. This can present an insuperable problem. For datasets where many of the values are less than 5 or any are equal to 0, it is necessary to substitute Fisher's exact test for the $\chi^2$-test (Section 10.7).

## 10.7. Fisher's exact test

Sir Ronald Aylmer Fisher (1890–1962) 'the father of modern statistics', developed the concept of likelihood:

> The likelihood of a parameter is proportional to the probability of the data and it gives a function which usually has a single maximum value, called the maximum likelihood.

He also contributed to the development of methods suitable for small samples and studied hypothesis testing. Fisher's exact test is an alternative to $\chi^2$ for testing the hypothesis that there is a statistically significant difference between two groups. It has the advantage that it does not make any approximations (Fisher's exact test), and so is suitable for small sample sizes. Fisher's exact test is a non-parametric test which assumes that:

1. The data analysed consist of nominal or ordinal variables.

2. The data consist of entire populations or be randomly sampled from the population, as in all significance tests.

3. The value of the first unit sampled has no effect on the value of the second unit – independent observations. Pooling data from

before–after tests or matched samples would violate this assumption.

4. A given case may fall in only one class – mutual exclusivity.

The formula for calculating Fisher's exact test is not complex, but can be tedious. Where:

$$
\begin{array}{ccc}
a & b & r_1 \\
c & d & r_2 \\
c_1 & c_2 & n
\end{array}
$$

$$P = (r_1!\ r_2!\ c_1!\ c_2!)/n!\ a!\ b!\ c!\ d!$$

As long as the criteria for test have been met, you can perform Fisher's test using statistics software or one of the many online calculators (search the internet for 'Fisher's' 'exact' and 'calculator').

## Problems (answers in Appendix 1)

**10.1.** The heights of a group of girls and a group of boys was measured. The frequency of measurements in both groups was found to have a normal distribution:

|                    | Girls   | Boys    |
|--------------------|---------|---------|
| Mean               | 1.25 m  | 1.29 m  |
| Standard deviation | 6 cm    | 5 cm    |

(a) Susan's height is 1.31 m. What is her $z$-score?

(b) Michael's height is 1.31 m. What is his $z$-score?

(c) Sally's $z$-score is $-1.2$. Is she taller or shorter than the average for her group?

(d) True or false: the boys' $z$-scores are higher than the girls' $z$-scores (explain your answer).

(e) What percentage of boys are taller than 1.39 m?

**10.2.** A group of 12 patients with high blood pressure is treated with drug A for 3 months. At the end of the treatment period, their blood pressure is measured and treatment with drug B started. After a further 3 months, their blood pressure is measured again. Analyse the data from this trial using Student's *t*-test:

|           | Drug A | Drug B |
|-----------|--------|--------|
| Patient 1  | 189 | 186 |
| Patient 2  | 181 | 181 |
| Patient 3  | 175 | 179 |
| Patient 4  | 186 | 189 |
| Patient 5  | 179 | 175 |
| Patient 6  | 191 | 189 |
| Patient 7  | 180 | 183 |
| Patient 8  | 183 | 181 |
| Patient 9  | 183 | 186 |
| Patient 10 | 189 | 190 |
| Patient 11 | 176 | 176 |
| Patient 12 | 186 | 183 |

(a) What sort of *t*-test should you perform to analyse these data?

(b) Should you use a one tailed or two-tailed test?

(c) How many degrees of freedom are there in this test?

(d) Is there a statistically significant difference at the 95% confidence level in the blood pressure of the patients after treatment with the two drugs?

**10.3.** In a study of the acidification of lakes, pH measurements were made of a series of lakes draining into two different rivers, A and B. Analyse the data from this trial using Student's *t*-test:

| A |      | B |      |
|------|------|------|------|
| 6.97 | 7.20 | 5.93 | 6.70 |
| 5.88 | 7.81 | 4.88 | 6.81 |
| 6.41 | 6.98 | 5.71 | 6.18 |
| 6.85 | 7.42 | 5.85 | 6.42 |
| 6.24 | 5.59 | 5.24 | 4.59 |
| 6.26 | 6.77 | 7.86 | 6.77 |
| 5.01 | 5.84 | 4.01 | 5.24 |
| 7.64 | 8.41 | 6.64 | 7.31 |
| 6.40 | 6.59 | 7.20 | 6.29 |
| 6.72 | 7.10 | 6.32 | 6.10 |

(a) What sort of $t$-test should you perform to analyse these data?

(b) Should you use a one-tailed or two-tailed test?

(c) How many degrees of freedom are there in this test?

(d) Is there a statistically significant difference at the 95% confidence level in the pH readings of the lakes draining into the two rivers?

**10.4.** The number of eggs in robins' nests in three different areas of woodland were counted and found to be:

A: 2, 0, 1, 1, 1, 3, 1, 3, 2, 1, 1, 2, 2, 2, 1, 3, 3, 1, 2, 0, 1, 1, 1, 1, 0

B: 2, 1, 2, 0, 1, 5, 1, 2, 3, 2, 1, 2, 2, 2, 0, 3, 2, 0, 1, 1, 0, 1, 0, 0, 1

C: 2, 0, 2, 0, 2, 5, 1, 2, 2, 1, 0, 1, 3, 2, 3, 2, 1, 1, 0, 1, 2, 1, 1, 4, 2

Can you perform an ANOVA test to demonstrate whether or not there a statistically significant difference at the 95% confidence level between the three woodlands?

**10.5.** A biologist measures the preference of three-spined sticklebacks (*Gasterosteus aculeatus*) for various food items. In a 3 h period, fish of length less than 4 cm consumed 14 *Daphnia galeata*, 14 *Daphnia magna* and 36 *Daphnia pulex*, while fish longer than 4 cm consumed 6 *Daphnia galeata*, 24 *Daphnia magna* and 31 *Daphnia pulex*. Use the $\chi^2$-test to compare the distribution of these variables and decide whether there is a statistically significant difference at the 95% confidence level between the feeding behaviour of the larger and the smaller sticklebacks.

(a) Construct a contingency table for the data.

(b) Formulate the null hypothesis for this experiment.

(c) How many degrees of freedom are there in this case?

(d) Calculate $\chi^2$.

(e) Is there a statistically significant difference at the 95% confidence level between the feeding behaviour of the larger and the smaller sticklebacks?

**10.6.** A group of 353 cancer patients are treated with a new drug. Of the patients who receive this treatment, 229 survive for more than 5 years after the commencement of treatment. Compare this result with a control group of 529 similar patients treated with the previously accepted drug therapy, 310 of whom survive for more than 5 years after the commencement of treatment. Is there a statistically significant difference at the 95% confidence level between the survival rates of the patients who received the new drug and those who received the previously accepted therapy?

# 11

# Correlation and Regression

---

**LEARNING OBJECTIVES:**

On completing this chapter, you should understand:

- the differences between correlation and regression;
- when to use each;
- the limitations of these tests.

---

## 11.1. Regression or correlation?

The correlation between two or more variables demonstrates the degree to which the variables are related. Linear regression demonstrates the relationship between selected values of $X$ and observed values of $Y$, from which the most probable value of $Y$ can be predicted for any value of $X$. Both correlation and regression are based on geometry and graphs and plots. Linear regression and correlation are similar and easily confused. In some situations it makes sense to perform both calculations. Calculate linear correlation if:

- You measured both $X$ and $Y$ in each subject and wish to quantify how well they are associated.

- Do not calculate a correlation coefficient if you manipulated both variables, for example salt intake (in diet) and blood pressure (by drug treatment). There is no point in measuring correlation unless one variable is independent (manipulated by experimentation) and the other dependent (on the first variable), or both variables are independent (from observation rather than experimentation).

Calculate linear regressions only if:

- One of the variables ($X$) is likely to precede or cause the other variable ($Y$).

- Choose linear regression if you manipulated the $X$ variable. It makes a big difference which variable is called $X$ and which is called $Y$, as linear regression calculations are not symmetrical with respect to $X$ and $Y$. If you swap the two variables, you will obtain a different regression line.

- In contrast, correlation calculations are symmetrical with respect to $X$ and $Y$. If you swap the labels $X$ and $Y$, you will still get the same correlation coefficient.

## 11.2. Correlation

Correlation, the relationship between two variables, is closely related to prediction. The greater the association between variables, the more accurately we can predict the outcome of events which have not yet happened. In biology, of all subjects, there is rarely an exact correlation of observed results with a mathematical function – the points never fit exactly on the line. The question is therefore whether an association between two variables could have occurred by chance. There are numerous methods for calculating correlation, for example:

1. The Pearson, or '$r$-value', correlation (parametric). Calculate the Pearson correlation coefficient if you know or can assume that both $X$ and $Y$ are interval data sampled from normally distributed populations. However, with large samples, the frequency distribution of the population is less important. Correlation is greatly affected by sample size. With very large samples small correlation coefficients will be statistically significant, but with small samples only very high correlations will reach statistical significance.

2. The Spearman (rank-order) correlation (non-parametric). Calculate the Spearman correlation coefficient if $X$ or $Y$ are ordinal or ranked

data. The Spearman correlation is based on ranking the two variables, and so makes no assumption about the distribution of the values and is usually calculated on occasions when it is not convenient, economic or even possible to give actual values to variables, but only to assign a rank order to instances of each variable. It may also be a better indicator that a relationship exists between two variables when the relationship between them is non-linear.

3. Many other methods (not described in this chapter), such as biserial correlation, tetrachoric correlation, etc.

A correlation coefficient is called $r$, for relationship. $r$-Values are dimensionless, that is they have no units. The $r$-value is a measure of the linear association between two variables which have been measured on interval or ratio scales, such as the relationship between height in metres and weight in kilograms. The correlation coefficient tells us the degree of linear association between the two variables, that is, how straight a line they form when plotted as a graph. However, $r$-values can be misleadingly small when there is a non-linear relationship between the variables. There are advanced correlation procedures not based on linear relationships but these are not covered in this chapter. Correlation analysis is performed in the same as any other statistical test of significance (Chapter 10):

1. Formulate the null hypothesis ('There is no proof of an association between the variables') and set the significance level ($\alpha$) before performing the test (e.g. 0.05).

2. Calculate the correlation coefficient ($r$) for the test data. The Pearson formula is fairly straightforward but rather cumbersome:

$$r = \frac{\sum XY - [(\sum X)(\sum Y)]/n}{\sqrt{\{\sum X^2 - [(\sum X)^2/n]\}\{\sum Y^2 - [(\sum Y)^2/n]\}}}$$

It is usually more convenient to use statistical software to perform the calculation (Appendix 2).

3. Determine whether the value of $r$ is equal to or greater than the critical value required to reject the null hypothesis. To do this you need to calculate the number of degrees of freedom for the test, i.e. take into account the number of independent observations used in the calculation of the test statistic. For a one-tailed test $df = n - 1$ and for a two-tailed test (most usual) $df = n - 2$.

4. Look up the calculated value of $r$ in a table of critical values of the correlation coefficient.

5. If the calculated value of $r$ is less than the critical value of $r$, accept the null hypothesis – there is no proof of an association between the variables. If the calculated value of $r$ is greater than or equal to the critical value of $r$, reject the null hypothesis – there is a significant association between the variables.

Calculation of the Spearman rank order correlation coefficient ($r_s$) is used when the data consists of ordinal variables (i.e. variables with an ordered series where numbers indicate rank order only). This is a non-parametric statistic but may be a better indicator than the Spearman coefficient that a relationship exists between two variables when the relationship between them is non-linear. If the data has not already been ranked, it must be converted into rank order before performing the Spearman calculation. When converting to ranks, the smallest value on X becomes a rank of 1, the second smallest becomes a rank of 2 and so on, for example:

| X | Y |   | X | Y |
|---|---|---|---|---|
| 7 | 4 |   | 2 | 1 |
| 5 | 7 | convert to ranks: | 1 | 2 |
| 8 | 9 |   | 3 | 4 |
| 9 | 8 |   | 4 | 3 |

The equation for the Spearman calculation is:

$$r_s = 1 - \frac{6 \sum D^2}{N(N^2 - 1)}$$

where $N$ is the number of pairs ($XY$) and $D$ is the difference between each pair ($X - Y$).

After calculating the value of $r_s$, this is compared with the critical value of $r$ as above in deciding whether to accept or reject the null hypothesis. Values of $r$ range from $+1$ (perfect correlation), through 0 (no correlation), to $-1$ (perfect negative correlation); see Figure 11.1.

In general terms, correlation coefficients up to 0.33 are considered to indicate weak relationships, between 0.34 and 0.66 indicate medium strength relationships and over 0.67 indicate strong relationships.

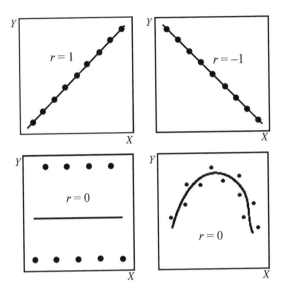

**Figure 11.1** Values of $r$ from $+1$ to $-1$

## Warning

Correlation tests are in some ways the most misused of all statistical procedures:

1. They are able to show whether two variables are connected. However, they are not able to prove that the variables are not connected, so do not overinterpret your results.

2. If one variable depends on another, i.e. there is a causal relationship so they are not independent, then it is always possible to find some kind of correlation between the two variables. However, if both variables depend on a third, they can show a strong correlation without any causal dependency between them, so take care. For example, there is a strong positive correlation between heart surgery and death rates, but does heart surgery cause the deaths? The missing third factor here is heart disease, which causes both heart surgery and deaths.

Scatter plots are useful in showing pictorially what the data to be analysed looks like. In particular, they suggest immediately whether there is a linear relationship between the datasets or not. They also reveal the degree of positive or negativeness of the association and show any extreme values which might be forcing linearity. Any coordinate which is far away from the others will be connected to the bulk of the datapoints by a straight line which might make the correlation appear significant. If the data are not linear (e.g. they form a curve or are bimodal), then the Pearson and Spearman methods cannot be used. Alternative non-linear regression methods such as polynomial regression should be used in these circumstances (but are not described in this chapter).

## Example

In patients undergoing renal (kidney) dialysis, is there a significant association between heart rate and blood pressure (see Table 11.1)? The scatter plot is shown in Figure 11.2. $H_0$ is that there is no association between

**Table 11.1**  Experimental data

| Patient | Heart rate | Blood pressure |
|---------|------------|----------------|
| 1 | 83 | 141 |
| 2 | 86 | 162 |
| 3 | 88 | 161 |
| 4 | 92 | 154 |
| 5 | 94 | 171 |
| 6 | 98 | 174 |
| 7 | 101 | 184 |
| 8 | 114 | 190 |
| 9 | 117 | 187 |
| 10 | 121 | 191 |

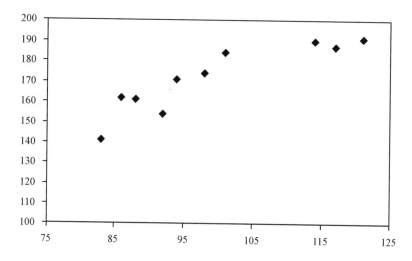

**Figure 11.2**  Scatter plot

the variables, $\alpha = 0.05$; Pearson $r$-value $= 0.903$; df $= 10 - 2 = 8$ (two-tailed); critical value of $r = 0.632$; $r_{\text{calc}}$ is greater than $r_{\text{crit}}$, so H$_0$ is rejected – there is evidence of a significant association between the heart rate and blood pressure in these patients. This can be seen visually by plotting a scatter graph of this data and drawing a trendline through it. *Warning*: you cannot accurately assess whether a significant correlation between variables exists by visual examination alone.

## 11.3. Regression

Regression and correlation are related but distinct statistical tests. Where correlation *quantifies* how closely two variables are connected, regression finds the line that best *predicts* Y from values of X. Simple linear regression aims to find a linear relationship between a response variable and a possible predictor variable by the method of *least squares*. Multiple linear regression aims to find a linear relationship between a response variable and several possible predictor variables. Non-linear regression aims to describe the relationship between a response variable and one or more explanatory variables in a non-linear fashion. Simple linear regression works by minimizing the sum of the square of the vertical distances of the points from the regression line, and hence is known as the 'least squares' method. The calculation effectively minimizes the sizes of squares drawn between the data points and the regression line (Figure 11.3).

The calculations involved in this process involve determining the 'residuals' between the actual and predicted values of Y. This is not

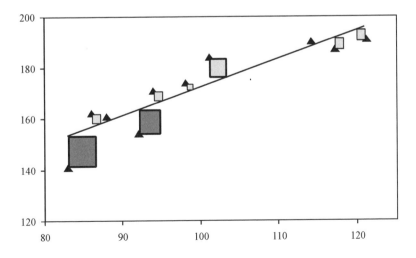

**Figure 11.3**  Simple linear regression

complex, but it is tedious and is almost always done using statistics software (Appendix 2). Plotting the residuals is a useful check of the regression model and can be done easily using software. A pattern in the residual plot can reveal flaws in the regression model, i.e. the data you are trying to fit do not plot as a straight line. Performing a regression analysis is similar to performing a correlation test:

1. Formulate the null hypothesis – 'Y is independent of X, therefore the slope of the regression line is 0'.

2. Calculate the test statistic. As with the correlation test, it is much more convenient to use statistical software to perform the calculation (Appendix 2).

3. Interpret the test statistic. The output of a linear regression analysis is an $r^2$-value. This is equivalent to the $r$-value from a correlation test, and shows how closely X and Y are related. By taking the square of the $r$-value, all values of $r^2$ become positive (remember that values of $r$ can range from $-1$ to $+1$), and fall between 0 (no correlation) and 1 (perfect correlation). The $r^2$-value tells you how much your ability to predict is improved by using the regression line, compared with not using it. The least possible improvement is 0, i.e. the regression line does not help at all. The greatest possible improvement is 1, i.e. the regression line fits the data perfectly. The value of $r^2$ is always between 0 and 1 since the regression line is never worse than worthless ($r^2 = 0$), and cannot be better than perfect ($r^2 = 1$). Unfortunately, $r$- and $r^2$-values only give a guide to the 'goodness-of-fit' and do not indicate whether an association between the variables is statistically significant. In order to do this, it is necessary to perform a statistical test of significance such as ANOVA or a $t$-test (Chapter 10). This is usually done automatically when performing linear regression with a statistics software and is another reason why it is much easier to use software for these calculations rather than performing them manually.

## *Example*

Using the data from the renal dialysis example in Section 11.2, we obtain an $r^2$-value of 0.815 (cf. a Pearson $r$-value of 0.903 in Section 11.2). Clearly this dataset shows a high degree of correlation between the two measurements, but what do the numbers mean exactly? An $r^2$-value is the square of the correlation coefficient. Multiplying $r^2$ by 100 gives the percentage of the variance of the two variables that is shared, e.g. an $r^2$-value of 0.5 means that $0.5 * 100 = 50\%$ of the variance of $y$ is 'explained' or predicted by the $x$ variable. However, it is important to remember that a very high value of $r^2$ can arise even though the relationship between the two variables is non-linear. The fit of a model should never simply be judged from the $r^2$-value alone – it is also necessary to formally test the statistical significance of the result, for example by comparing the calculated value of the statistic with the critical value.

## Problems (answers in Appendix 1)

**11.1.** Construct scatter plots of the four datasets in Table 11.2 and state whether each is suitable for correlation analysis or not.

Table 11.2   Datasets A–D

| A | | B | | C | | D | |
|---|---|---|---|---|---|---|---|
| X | Y | X | Y | X | Y | X | Y |
| 56 | 23 | 0.50 | 0.13 | 1.00 | 1.006 | 0.310 | 0.503 |
| 51 | 37 | 1.00 | 0.11 | 1.00 | 2.019 | 0.754 | 0.214 |
| 5 | 14 | 1.50 | 0.92 | 1.00 | 3.019 | 0.570 | 0.680 |
| 70 | 5 | 2.00 | 0.41 | 1.00 | 4.016 | 0.273 | 0.462 |
| 55 | 65 | 2.50 | 1.45 | 1.00 | 5.003 | 0.162 | 0.568 |
| 38 | 64 | 3.00 | 1.37 | 1.00 | 6.014 | 0.069 | 0.867 |
| 66 | 32 | 3.50 | 0.20 | 1.00 | 7.003 | 0.854 | 0.528 |
| 25 | 64 | 4.00 | 2.33 | 1.00 | 8.009 | 0.372 | 0.809 |
| 17 | 34 | 4.50 | 2.98 | 1.00 | 9.009 | 0.751 | 0.271 |
| 59 | 31 | 5.00 | 1.57 | 1.00 | 10.004 | 0.128 | 0.741 |
| 1 | 16 | 5.50 | 1.75 | 1.00 | 11.006 | 0.679 | 0.123 |
| 17 | 8 | 6.00 | 3.63 | 1.00 | 12.008 | 0.515 | 0.170 |
| 19 | 20 | 6.50 | 3.87 | 1.00 | 13.000 | 0.978 | 0.103 |
| 23 | 46 | 7.00 | 0.90 | 1.00 | 14.007 | 0.354 | 0.216 |
| 1 | 42 | 7.50 | 4.18 | 1.00 | 15.016 | 0.561 | 0.445 |
| 5 | 7 | 8.00 | 1.31 | 1.00 | 16.007 | 0.379 | 0.217 |
| 60 | 36 | 8.50 | 1.32 | 1.00 | 17.014 | 0.355 | 0.796 |
| 13 | 45 | 9.00 | 3.65 | 1.00 | 18.006 | 0.196 | 0.744 |
| 49 | 76 | | | 1.00 | 19.001 | 0.077 | 0.951 |
| 76 | 7 | | | 1.00 | 20.011 | | |

**11.2.** In a biology practical, the lengths and weights of 10 earthworms are measured (Table 11.3). Determine the Pearson correlation coefficient for this data. What can you say about the relationship between the length and the weight of an earthworm?

Table 11.3   Lengths and weights of 10 earthworms

| Sample | Length (cm) | Weight (g) |
|--------|-------------|------------|
| 1 | 5.37 | 5.38 |
| 2 | 3.99 | 4.30 |
| 3 | 5.11 | 6.35 |
| 4 | 4.54 | 6.20 |
| 5 | 2.44 | 3.12 |
| 6 | 6.42 | 8.04 |
| 7 | 7.26 | 8.35 |
| 8 | 2.60 | 3.31 |
| 9 | 3.32 | 4.33 |
| 10 | 6.54 | 7.97 |

**11.3.** The lengths of the femur and humerus were measured and compared in 12 human skeletons. The data are presented here in rank order only, where 1 = shortest, 12 = longest (Table 11.4). Is there a statistically significant correlation between the length of the two bones?

Table 11.4   Lengths of femur and humerus

| Subject | Length of femur (rank) | Length of humerus (rank) |
|---------|------------------------|--------------------------|
| 1 | 3 | 2 |
| 2 | 5 | 10 |
| 3 | 2 | 5 |
| 4 | 11 | 11 |
| 5 | 4 | 3 |
| 6 | 6 | 6 |
| 7 | 10 | 9 |
| 8 | 9 | 8 |
| 9 | 8 | 7 |
| 10 | 1 | 1 |
| 11 | 7 | 4 |
| 12 | 12 | 12 |

**11.4.** The reaction rate of the enzyme lactate dehydrogenase was measured at different concentrations of the substrate, pyruvate (Table 11.5). Using statistics software (Appendix 2) the $r^2$-value for this experiment was determined as 0.939. How would you describe the accuracy of the results of the experiment?

Table 11.5  Reaction rate of the enzyme lactate dehydrogenase

| Pyruvate ($\mu$M) | Reaction rate ($\mu$mol s$^{-1}$) |
| --- | --- |
| 25 | 0.0056 |
| 50 | 0.0098 |
| 75 | 0.0144 |
| 100 | 0.0149 |
| 125 | 0.0175 |
| 150 | 0.0248 |
| 175 | 0.0243 |
| 200 | 0.0326 |
| 225 | 0.0306 |
| 250 | 0.0367 |
| 275 | 0.0326 |
| 300 | 0.0361 |

**11.5.** What does an $r^2$-value of 0.59 mean? Explain.

# Appendix 1: Answers to Problems

## Chapter 2

### Basic equations

**2.1.** Solve for $x$ (find the value of $x$):

$$3x + 3 = 5$$
$$3x + 3 - 3 = 5 - 3$$
$$3x = 2$$
$$3x/3 = 2/3$$
$$x = 2/3$$

*Check answer*

$$3(2/3) + 3 = 5$$

**2.2.** Solve for $y$:

$$5y + 12 = 22$$
$$5y + 12 - 12 = 22 - 12$$
$$5y = 10$$
$$5y/5 = 10/5$$
$$y = 2$$

*Check answer*

$$5(2) + 12 = 22$$

**2.3.** Solve for $z$:

$$10z + 9 = 6$$
$$10z + 9 - 9 = 6 - 9$$
$$10z = -3$$
$$10z/10 = -3/10$$
$$z = -3/10$$

## Check answer

$$10(-3/10) + 9 = 6$$

**2.4.** Solve for $w$:

$$8w + 8 = 12$$
$$8w + 8 - 8 = 12 - 8$$
$$8w = 4$$
$$8w/8 = 4/8$$
$$w = 4/8$$
$$w = 1/2$$

## Check answer

$$8(1/2) + 8 = 12$$

**2.5.** Solve for $h$:

$$9h + 86 = 99$$
$$9h + 86 - 86 = 99 - 86$$
$$9h = 13$$
$$h = 13/9$$

## Check answer

$$9(13/9) + 86 = 99$$

**2.6.** Solve for $a$:

$$77a - 75 = 1$$
$$77a - 75 + 75 = 1 + 75$$
$$77a = 76$$
$$a = 76/77$$

## Check answer

$$77(76/77) - 75 = 1$$

**2.7.** Solve for $B$:

$$11B + 11 = 11$$
$$11B + 11 - 11 = 11 - 11$$
$$B = 0$$

## Check answer

$$11(0) + 11 = 11$$

**2.8.** Solve for $\phi$:

$$123\phi = 1353 = -123$$
$$123\phi - 1353 + 1353 = -123 + 1353$$
$$123\phi = 1230$$
$$123\phi/123 = 1230/123$$
$$\phi = 10$$

## Check answer

$$123(10) - 1353 = -123$$

**2.9.** Solve for $f$:

$$4f + 12 = 17$$
$$4f + 12 - 12 = 17 - 12$$
$$4f = 5$$
$$f = 5/4$$

## Check answer

$$4(5/4) + 12 = 17$$

**2.10.** Solve for $x$:

$$2x = 3x - 2$$
Subtract $3x$ from each side:
$$2x - 3x = 3x - 3x - 2$$
Simplify:
$$-x = -2$$
Divide each side by $-1$:
$$x = 2$$

## Check answer

$$2(2) = 3(2) - 2$$

**2.11.** Solve for $x$:

$4x = 2x - 3$
Subtract $2x$ from each side:
$4x - 2x = 2x - 2x - 3$
$2x = -3$
Divide each side by 2:
$x = -3/2$

## Check answer

$4(-3/2) = 2(-3/2) - 3$

**2.12.** Solve for $p$:

$3p = p + 6$
Subtract $p$ from each side:
$2p = 6$
Divide each side by 2:
$p = 3$

## Check answer

$3(3) = 3 + 6$

**2.13.** Solve for $z$:

$4z = 2z - 5$
Subtract $2z$ from each side:
$2z = -5$
Divide each side by 2:
$z = -5/2$

## Check answer

$4(-5/2) = 2(-5/2) - 5$

**2.14.** Solve for $a$:

$22a = 41a - 38$
Subtract $41a$ from each side:
$-19a = -38$
Divide each side by $-19$:
$a = -38/-19$
Simplify $a = 2$

## Check answer

$$22(2) = 41(2) - 38$$

**2.15.** Solve for $x$:

$(x/6) = (x/2) + (5/4)$
Find a common denominator (lowest common multiple of all the denominators):
$(2x/12) = (6x/12) + (15/12)$
Multiply both sides by 12:
$2x = 6x + 15$
Subtract $6x$ from each side:
$-4x = 15$
Divide each side by $-4$:
$x = -15/4$

## Check answer

$$[(-15/4)/6] = [(-15/4)/2] + (5/4)$$

**2.16.** Solve for $t$:

$(t/3) = (t/6) + (1/3)$
Find a common denominator:
$(2t/6) = (t/6) + (2/6)$
Multiply each side by 6:
$2t = t + 2$
Subtract $t$ from each side:
$t = 2$

## Check answer

$$(2/3) = (2/6) + (1/3)$$

**2.17.** Solve for $w$:

$(2w/3) + 3 = (w/4)$
Find a common denominator and convert 3 to a fraction:
$(8w/12) + (36/12) = (3w/12)$
Multiply each side by 12:
$8w + 36 = 3w$
Subtract $8w$ from each side:
$36 = -5w$
Divide each side by $-5$:
$-36/5 = w$

*Check answer*

$$[2(-36/5)/3]+3=[(-36/5)/4]$$

**2.18.** Solve for $m$:

$(100m/3)+(22/33)=(101m/3)$
Find a common denominator:
$(1100m/33)+(22/33)=(1111m/33)$
Multiply both sides by 33:
$1100m+22=1111m$
Subtract $1100m$ from both sides:
$22=11m$
Divide both sides by 11:
$22/11=m$
$m=2$

*Check answer*

$$[100(2)/3]+(22/33)=[101(2)/3]$$

## Multiple variable equations

**2.19.** Solve for $x$ (find the value of $x$):

$x+9=y$
Subtract 9 from each side:
$x=y-9$

*Check answer*

$$y-9+9=y$$

**2.20.** Solve for $B$:

$10B+2=z-6$
Subtract 2 from each side:
$10B=z-8$
Divide each side by 10:
$B=z/10-4/5$

*Check answer*

$$10(z/10-4/5)+2=z-6$$

**2.21.** Solve for $n$:

$3n + 6 = x - 10$
Subtract 6 from each side:
$3n = x - 16$
Divide each side by 3:
$n = (x - 16)/3$

## Check answer

$3[(x - 16)/3] + 6 = x - 10$

**2.22.** Solve for $y$:

$2y + 3 = 2x + 3$
Subtract 3 from each side:
$2y = 2x$
Divide each side by 2:
$y = x$

## Check answer

$2x + 3 = 2x + 3$

**2.23.** Solve for $y$:

$5y + 6 = 2x$
Subtract 6 from each side:
$5y = 2x - 6$
Divide each side by 5:
$y = (2x - 6)/5$

## Check answer

$5[(2x - 6)/5] + 6 = 2x$

**2.24.** Solve for $c$:

$c - 8 = 4z + 2$
Add 8 to each side:
$c = 4z + 10$

## Check answer

$4z + 10 - 8 = 4z + 2$

## Word problems

**2.25.** The sum of nine plus twice a number equals twenty-three. What is the value of the number?

Write the problem as an equation:
$9 + 2x = 23$
Subtract 9 from each side:
$2x = 14$
Divide each side by 2:
$x = 7$

### Check answer

$9 + 2(7) = 23$

**2.26.** Forty-four mealworms are placed in an escape-proof bowl in a lizard's cage. In twenty-four hours, the lizard visits the bowl to feed three times, eating the same number of mealworms each time. At the end of the experiment, eight mealworms remain uneaten. How many mealworms does the lizard eat on each visit to the food bowl? (Hint: the difference between forty-four and three times a number is equal to eight.)

Write the problem as an equation:
$44 - 3x = 8$
Subtract 44 from each side:
$-3x = -36$
Divide each side by $-3$:
$x = 12$

### Check answer

$44 - 3(12) = 8$

**2.27.** The spine of a mammal contains forty-three vertebrae, the same number in each of the cervical, thoracic, lumbar, sacral and coccygeal regions, plus an additional three vertebrae. How many vertebrae are present in each region of the spine of this species?

Write the problem as an equation:
$3 + 5n = 43$
Subtract 3 from each side:
$5n = 40$
Divide each side by 5:
$n = 8$

## Check answer

$$3 + 5(8) = 43$$

**2.28.** A total of nine results are available from an experiment performed by students. Six students performed the experiment on Tuesday and each obtained a result. The rest of the class performed the experiment on Wednesday, but only one-third of these students obtained a result. How many students performed the experiment on Wednesday?

Write the problem as an equation:
$6 + x/3 = 9$
Subtract 6 from each side:
$x/3 = 3$
Multiply each side by 3:
$x = 9$

## Check answer

$$6 + x/3 = 9$$

**2.29.** Nine birds of prey raised as part of a captive breeding programme are tagged with radio transmitters and released into the wild. One year later, only four birds are still alive. How many of the birds released have died?

Write the problem as an equation:
$9 - n = 4$
Subtract 9 from each side:
$-n = -5$
Multiply each side by $-1$ (to change sign):
$n = 5$

## Check answer

$$9 - 5 = 4$$

**2.30.** Sheila starts the term with £240 in her bank account. Each week she withdraws the same amount of money. After six weeks, her account is £90 overdrawn. How much money did she withdraw each week?

Write the problem as an equation:
$240 - 6x = -90$
Subtract 240 from both sides:
$-6x = -330$
Divide each side by $-6$:
$x = 330/6( = £55)$

## Check answer

$$240 - 6(55) = -90$$

**2.31.** A researcher counts eleven male robins with successful breeding territories in an area of $2400\,m^2$. Then they observe another successful breeding male in the same area. What is the average area occupied by each breeding male?

Write the problem as an equation:
$n + 11n = 2400$
Combine like terms:
$12n = 2400$
Divide each side by 12:
$n = 200\ (m^2)$

## Check answer

$$200 + 11(200) = 2400$$

**2.32.** Nine students go out for a meal. In the restaurant each student puts the same amount of money on the table to pay for the meal. The bill is £142 and £2 is left over. How much money did each student put on the table?

Write the problem as an equation:
$9x - 142 = 2$
Add 142 to each side:
$9x = 144$
Divide each side by 9:
$x = £16$

## Check answer

$$9(16) - 142 = 2$$

**2.33.** At pH 7.4 an enzyme transforms 45 mmol of substrate $min^{-1}$. At the suboptimal pH of 6.9, the enzyme transforms 12 mmol of substrate $min^{-1}$ less than at pH 7.4. How much substrate is transformed per minute at pH 6.9?

Write the problem as an equation:
$n + 12 = 45$
Subtract 12 from each side:
$n = 33\ (mmol\ min^{-1})$

## Check answer

$$33 + 12 = 45$$

## Fractions

**2.34.** Calculate: $\frac{3}{2} + \frac{2}{3}$ etc

Find a common denominator:
$\frac{9}{6} + \frac{4}{6} = \frac{13}{6}$

**2.35.** Calculate: $\frac{7}{9} - \frac{5}{6}$

Find a common denominator:
$\frac{14}{18} - \frac{15}{18} = \frac{-1}{18}$

**2.36.** Calculate: $\frac{8}{9} * \frac{5}{7}$

$= \frac{40}{63}$

**2.37.** Calculate: $\frac{6}{7} / \frac{5}{6}$

Flip (take the reciprocal of) the divisor and multiply:
$\frac{6}{7} * \frac{6}{5} = \frac{36}{35}$

**2.38.** Express as a fraction: $1 + \frac{6}{8}$

$= \frac{8}{8} + \frac{6}{8} = \frac{14}{8}$

**2.39.** Express as a fraction: $2 + \frac{1}{12}$

$= \frac{24}{12} + \frac{1}{12} = \frac{25}{12}$

**2.40.** Express as a fraction: $2 + \frac{8}{11}$

$= \frac{22}{11} + \frac{8}{11} = \frac{30}{11}$

## Ratios

**2.41.** Are 4:12 and 36:72 equal ratios?

$4/12 = 0.33$
$36/72 = 0.5$
The quotients are not equal so these are not equal ratios.

**2.42.** Seventy-five per cent of the prey items captured by bats are moths. Express the amount of moths in a bat's diet as a ratio.

The bat eats 75% moths:25% other prey $= 3:1$

**2.43.** If $x = 6$ and the ratio of $x:y = 2:5$, what is the value of $y$?

$$\frac{x}{y} = \frac{2}{5}$$

So

$$\frac{6}{y} = \frac{2}{5}$$
$$y * 2 = 6 * 5$$
$$2y = 30$$
$$y = 15$$

**2.44.** If a swallow flies 100 metres in 10 seconds, how long would it take it to fly 15 400 metres?

$$\frac{100\,\text{m}}{10\,\text{s}} = \frac{15\,400\,\text{m}}{x\,\text{s}}$$
$$100x = 154\,000\,\text{s}$$
$$1x = 1540\,\text{s}$$

$$1540/60 = 25.67\,\text{minutes} \ (24\,\text{minutes}\ 40\,\text{seconds})$$

**2.45.** What volume of a $5\,\text{gL}^{-1}$ solution is required to make 100 mL of a $2\,\text{gL}^{-1}$ solution?

The ratio of the concentrations of the two solutions is $5:2 = 2.5$.
The volume of solution required is 100 mL, so:
$100/2.5 = 40\,\text{mL}$ of $5\,\text{gL}^{-1}$ solution (plus 60 mL of solvent).
(See Chapter 4 for more problems of this type.)

# Chapter 3

## True or false?

**3.1.** The litre (L) is the SI unit of volume.

False. The cubic metre $(\text{m}^3)$ is SI unit of volume. The litre is a metric unit but not an SI unit.

**3.2.** The hertz (Hz) is the SI unit of frequency.

True. $1\,\text{Hz} = 1$ oscillation per second $(1\,\text{s}^{-1})$.

**3.3.** The kilogram (kg) is the SI unit of weight.

False. The kilogram is the SI unit of mass.

**3.4.** The Celsius is the SI unit of temperature.

False. The kelvin (K) is the SI unit of temperature.

**3.5.** The newton (N) is the SI unit of power.

False. The newton is the SI unit of force. The SI unit of power is the watt (W).

## Are the following units written correctly or wrongly?

**3.6.** 3 Kg

Wrong – the kilo $(1 * 10^3)$ prefix is written as a lower case k, so 3 kg is correct.

**3.7.** The acceleration due to gravity is $9.8\,m\,s^{-2}$.

Wrong – there should be a space between the number and the units, so $9.8\,m\,s^{-2}$ is correct.

**3.8.** The power output of the human heart is about 5 Watts.

Wrong – it should be 5 watts (5 W if written as an abbreviation).

**3.9.** A virus particle is 25 μmm in diameter.

Wrong – this is a good example of how writing SI units wrongly can be ambiguous. 25 μmm could be interpreted in several ways, but if it is read as '25 micro-milli-metres' it should be written as 25 nanometres (25 nm = $25 * 10^{-9}$ m).

**3.10.** The output of a nuclear power station is 23 GW.

Correct (23 gigawatts).

## Converting SI units

**3.11.** Write 15 mm as nm.

$15\,mm = 15 * 10^{-3}\,m$. $1\,nm = 10^{-9}\,m$, so $15\,mm = 15 * 10^6\,nm$.

**3.12.** Write 3 Pa as μPa.

$1\,Pa = 1 * 10^6\,μPa$, so $3\,Pa = 3 * 10^6\,μPa$.

**3.13.** Write $14 * 10^9\,g$ as kg.

$1\,kg = 1 * 10^3\,g$. $14 * 10^9 / 1 * 10^3 = 14 * 10^6\,kg$.

**3.14.** Write $1\,m^3$ as $cm^3$.

$1\,m = 100\,cm$. $1\,m^3 = 100 * 100 * 100 = 1\,000\,000\,cm^3$ $(= 1 * 10^6\,cm^3)$.

**3.15.** Write $1200\,pg$ as $ng$.

$1200\,pg = 1200 * 10^{-12}\,g = 1.2 * 10^{-9}\,g = 1.2\,ng$.

**3.16.** What is the equivalent of $100°C$ in K?

$K = 273 + °C$. $273 + 100 = 373\,K$.

**3.17.** What is the equivalent of $274\,K$ in $°C$?

$°C = K - 273 = 274 - 273 = 1°C$.

**3.18.** Write $0.005\,kg\,cm^{-3}$ $g\,m^{-3}$.

$0.005\,kg\,cm^{-3} = 5\,g\,cm^{-3}$. $1\,m^3 = 1 * 10^6\,cm^3$, so $5\,g\,cm^{-3} = 5 * 10^6\,g\,m^{-3}$.

**3.19.** Convert $15\,m\,s^{-1}$ into $km\,h^{-1}$.

There are 60 seconds in one minute and 60 minutes in one hour, so $60 * 60 = 3600\,s * 15 = 54\,000\,m\,h^{-1} = 54\,km\,h^{-1}$.

**3.20.** Convert $29\,m^3$ into L.

$1\,m^3 = 1 * 10^6\,cm^3$ (see question 3.14). $1\,L = 1 * 10^3\,cm^3$, so $1\,m^3 = 1 * 10^3\,L$. $29\,m^3 = 29 * 10^3\,L$ ($29\,000\,L$).

## Energy

**3.21.** What force is exerted in a chair by a $70\,kg$ person standing on it?

$$f = m * a = kg * m\,s^{-2}$$
$$= 70 * 9.8\,m\,s^{-2} = 686\,N$$

**3.22.** A contracting muscle fibre exerts a force of $1\,pN$ and moves the anchor point of the fibre $1\,nm$. How much work does the fibre do?

$$work = force * distance = N * m$$
$$= 1 * 10^{-12}\,N * 1 * 10^{-9}\,m = 1 * 10^{-21}\,J$$

**3.23.** If the muscle fibre contracts in $55\,ms$, what power does it exert?

$$power = work/time = J/s$$
$$= 1 * 10^{-21}\,J/0.055\,s = 1.8 * 10^{-20}\,W$$

**3.24.** What is the weight of a seashell with a mass of 48 g lying on a beach?

$$\text{weight} = \text{mass} * \text{acceleration due to gravity at sea level}$$
$$= 0.048\,\text{kg} * 9.8\,\text{m}\,\text{s}^{-2} = 0.47\,\text{N}$$

**3.25.** What is the gravitational potential energy of a 2.1 kg coconut hanging from a tree 3.4 m above a beach?

$$\text{gravitational potential energy} = \text{weight} * \text{height} = \text{N} * \text{m}$$
$$\text{mass} = \text{weight}/\text{acceleration due to gravity} = 2.1\,\text{kg}/9.8\ \text{m}\,\text{s}^{-2} = 0.21\,\text{N}$$
$$0.21\,\text{N} * 3.4\,\text{m} = 0.714\,\text{J}$$

**3.26.** What is the kinetic energy of a 2.1 kg coconut which strikes the head of a sunbather at a speed of $3.3\,\text{m}\,\text{s}^{-2}$?

$$\text{kinetic energy} = 0.5 * \text{mass} * \text{speed}^2$$
$$\text{mass} = \text{weight}/\text{acceleration due to gravity} = 2.1\,\text{kg}/9.8\,\text{m}\,\text{s}^{-2} = 0.21\,\text{N}$$
$$\text{kinetic energy} = 0.5 * 0.21\,\text{N} * 3.3^2 = 1.1\,\text{J}$$

**3.27.** An electrophoresis gel carrying a current of 38 mA has a resistance of $5.25 * 10^4\,\text{W}$. Calculate the voltage across the gel.

$$\text{potential} = \text{current} * \text{resistance}$$
$$= 0.038 * 5.25 * 10^4 = 1995\,\text{V}\ (1.995\,\text{kV})$$

**3.28.** An electrophoresis gel has a resistance of $4.33 * 10^4\,\text{W}$ and a potential of 1920 V across the gel. What current is passing through the gel?

$$\text{current} = \text{potential}/\text{resistance}$$
$$= 1920/4.33 * 10^4 = 44.34\,\text{mA}$$

**3.29.** An electrophoresis gel has a potential of 1733 V and conducts a current of 55 mA. Calculate the electrical resistance of the gel?

$$\text{resistance} = \text{potential}/\text{current}$$
$$= 1733/0.055 = 31\,509\,\Omega$$

**3.30.** A water sample has a resistance of $6505\,\text{W}\,\text{cm}^{-1}$. Calculate the conductivity of the sample.

$$\text{conductivity} = 1/\text{resistance}$$
$$= 1/6505 = 1.537 * 10^{-4}\,\text{S}\,\text{cm}^{-1}\ (153.7\,\mu\text{S}\,\text{cm}^{-1})$$

# Chapter 4

**4.1.** One-hundred and twenty grams of NaOH are dissolved in water to make 5440 mL of solution. The molecular weight of NaOH is 40. What is the molarity of the resulting solution?

Amount of NaOH:

$$120/40 = 3\,\text{mol}$$

Concentration (mol/L = molarity):

$$3/5.44 = 0.55\,\text{M}$$

**4.2.** How many grams of NaCl are needed to make 120 mL of a 0.75 M solution? The molecular weight of NaCl is 58.44.

Concentration of NaOH:

$$58.44 * 0.75 = 43.83\,\text{g}\,\text{L}^{-1}$$

Amount:

$$43.83 * 0.12 = 5.26\,\text{g}$$

**4.3.** Sea water contains roughly 28.0 g of NaCl per litre. The molecular weight of NaCl is 58.44. What is the molarity of NaCl in sea water?

Moles/volume (L) = Molarity:

$$28.0/58.44 = 0.48\,\text{mol}\,\text{L}^{-1} = 0.48\,\text{M}$$

**4.4.** One-hundred and twenty-seven grams of NaCl and 19.9 g of sodium azide (NaN$_3$) are dissolved in water to make a 55 mL solution. The molecular weight of NaCl is 58.44 and that of NaN$_3$ is 65.01. What is the molarity of NaN$_3$ in the solution?

Does the presence of NaCl influence the molarity of NaN$_3$? No, so, amount of NaN$_3$:

$$19.9/65.01 = 0.306\,\text{mol}$$

Concentration:

$$0.306/0.055 = 5.57\,\text{M}$$

**4.5.** Undiluted sulphuric acid (H$_2$SO$_4$, molecular weight 98.07) is a 98% (w/v) solution. What is the molarity of the undiluted solution?

1 L of 98% (w/v) solution contain 980 g of solute.

$$\text{grams}/M_w = \text{mol}$$

$980/98.07 = 9.99\,\text{mol in }1\,\text{L} = 9.99\,\text{M}.$

**4.6.** The molecular weight of bovine serum albumin (BSA) is 66 200. How many moles of BSA are present in 15 mL of 50 mg mL$^{-1}$ BSA solution?

15 mL of solution contains $15 * (5 * 10^{-2})\,\text{g BSA}, = 0.75\,\text{g}.$

$$0.75/66\,200 = 1.13 * 10^{-5}\,\text{mol in}15\,\text{mL}$$

so:

$$1000/15 * 1.13 * 10^{-5}\,\text{mol} = 7.53 * 10^{-4}\,\text{mol in}1\,\text{L}, = 7.53 * 10^{-4}\,\text{M}$$

**4.7.** What volume of 0.9 M KCl is needed to make 225 mL of 0.11 M solution? The molecular weight of KCl is 74.55.

This is a dilution question – the molecular weight of KCl is irrelevant.

$$M_1 V_1 = M_2 V_2$$

$0.9 * V_1 = 0.11 * 225.\ V_1 = (0.11 * 225)/0.9 = 27.5\,\text{mL}$

An easier way? Work out the ratio of the molarities:

$$0.11/0.9 = 1.22.\ 1.22 * 225 = 27.5\,\text{mL}.$$

**4.8.** 'TE' is a frequently used buffer solution for DNA and contains 10 mM Tris–HCl pH 7.5, 1 mM EDTA. You have a 1 M stock solution of Tris–HCl pH 7.5 and a 0.5 M stock solution of EDTA. What volume of each stock solution do you need to make 333 mL of TE buffer?

  (a) Tris:

$M_1 V_1 = M_2 V_2.\ 1.0 * V_1 = 0.01 * 333.\ V_1 = (0.01 * 333)/1.0 = 3.33\,\text{mL}$

  (b) EDTA – use the ratio method for this one:

$$0.001/0.5 = 0.002 * 333 = 0.66\,\text{mL}$$

**4.9.** Chloramphenicol is soluble in ethanol at 0.1 g mL$^{-1}$ but much less soluble in water. What volume of 0.1 g mL$^{-1}$ solution must be added to 100 mL of a bacterial culture to give a final concentration of 150 µg mL$^{-1}$?

To avoid confusion, standardize the units. For this calculation, work in g mL$^{-1}$:

$$0.1\,\text{g mL}^{-1}/1.5 * 10^{-4}\,\text{g mL}^{-1} = 666.67$$

i.e. the stock solution is 666.67 times more concentrated than the desired end result, so:

$$100/666.67 = 0.15\,\text{mL}.$$

**4.10.** Urea lysis buffer contains the following ingredients:

9.9 g urea ($M_w$ 60.06)/100 mL

22 g SDS ($M_w$ 288.4)/100 mL

77 mL 5 M NaCl stock solution/100 mL

2.5 mL 0.2 M EDTA stock solution/100 mL

15 mL 1 M Tris–HCl pH 8.0 stock solution/100 mL

What are the final molar concentrations of EACH of the components of this buffer?

(a) Urea concentration:

9.9/60.06 = 0.165 mol in 100 mL, therefore 1.65 mol in 1000 mL = 1.65 M.

(b) SDS concentration:

22/288.4 = 0.076 mol in 100 mL, therefore 0.76 mol in 1000 mL = 0.76 M.

(c) Amount of NaCl:

$$5(\text{mol L}^{-1}) * 0.077\ (\text{L}) = 0.385\,\text{mol}$$

Concentration:

0.385 mol in 100 mL, therefore 3.85 mol in 1000 mL = 3.85 M.

(d) Amount of EDTA:

$$0.2(\text{mol L}^{-1}) * 0.0025\ (\text{L}) = 0.0005\,\text{mol}$$

Concentration:

0.0005 mol in 100 mL, therefore 0.005 mol in 1000 mL = 0.005 M (= 5 mM).

(e) Amount of Tris:

$$1(\text{mol L}^{-1}) * 0.015\ (\text{L}) = 0.015\,\text{mol}$$

Concentration:

0.015 mol in 100 mL, therefore 0.15 mol in 1000 mL = 0.15 M (= 150 mM).

**4.11.** A solution of DNA with a concentration of $50\,\mu g\,mL^{-1}$ has an $A_{260}$ of 1.0. What is the concentration of the following solutions:

(a) $A_{260}$ 0.65

Concentration:

$$50\,\mu g\,mL^{-1} * 0.65 = 32.5\,\mu g\,mL^{-1}$$

(b) $A_{260}$ 0.31 after diluting 15-fold

Concentration:

$$50\,\mu g\,mL^{-1} * 0.31 * 15 = 232.5\,\mu g\,mL^{-1}.$$

**4.12.** A solution of RNA with a concentration of $40\,\mu g\,mL^{-1}$ has an $A_{260}$ of 1.0. What is the concentration of the following solutions:

(a) $A_{260}$ 0.59

Concentration:

$$40\,\mu g\,mL^{-1} * 0.59 = 23.6\,\mu g\,mL^{-1}$$

(b) $A_{260}$ 0.48 after diluting 10-fold

Concentration:

$$40\,\mu g\,mL^{-1} * 0.48 * 10 = 192\,\mu g\,mL^{-1}$$

**4.13.** The amino acid tyrosine has an extinction coefficient of $1405\,M^{-1}\,cm^{-1}$ at 274 nm. Calculate:

(a) The $A_{274}$ of a 0.4 mM solution of tyrosine measured in a cuvette with a 1 cm light path.

$$A_{274} = 1405\,M^{-1}\,cm^{-1} * 0.0004 = 0.562$$

(b) The concentration of a solution of tyrosine with $A_{274}$ 0.865 after 12-fold dilution measured in a cuvette with a 1 cm light path.

Concentration:

$$A/(e * 1) = [0.865/(1405 * 1)] * 12 = 0.0074\,M = 7.4\,mM$$

**4.14.** A sample of a culture of bacteria is subjected to 10-fold serial dilution; 0.1 mL aliquots of the dilutions are grown on agar plates and the number of colonies counted:

| Dilution: | $10^{-1}$ | $10^{-2}$ | $10^{-3}$ | $10^{-4}$ | $10^{-5}$ | $10^{-6}$ |
|---|---|---|---|---|---|---|
| Number of colonies: | Too many to count | Too many to count | 249 | 24 | 2 | 0 |

Assuming that one cell gives rise to one colony, how many cells were there per mL of the original culture?

For maximum accuracy, it is important to pick the most appropriate sample – the one with the largest number of colonies which can be counted accurately.

$$N = \frac{R * D}{V}$$

$$= (249 * 10^3)/0.1 \, \text{mL} = 2.49 * 10^6 \, \text{cells mL}^{-1}$$

**4.15.** A virus suspension was serially diluted to perform a plaque assay:

Dilution A was 0.1 mL + 4.9 mL

Dilution B was 1 mL of A + 1 mL

Dilution C was 0.01 mL of B + 0.99 mL

Dilution D was 0.1 mL of C + 9.9 mL

(a) Calculate the final dilution:

$$A = 0.1 + 4.9 = 5 \quad B = 1 + 1 = 2 \quad C = 0.01 + 0.99 = 100$$
$$D = 0.1 + 9.9 = 10$$
$$5 * 2 * 100 * 10 = 10\,000, \text{i.e. a} 1{:}10\,000 \text{ dilution}$$

(b) Calculate the reciprocal of the final dilution.

$$1/1 * 10^{-4} = 10\,000$$

(c) If the original suspension contained 12 000 000 virus particles mL$^{-1}$ and 0.1 mL of each dilution was used in the plaque assay, which dilution should be used to determine the density of the original suspension?

$$N = \frac{R * D}{V} \quad \text{so} \quad R = \frac{N * V}{D}$$

Assay of dilution A will give: $12 * 10^6 * 0.1 * (1/5) = 240\,000$ plaques, too many to count.

Assay of dilution B will give: $12 * 10^6 * 0.1 * (1/10) = 120\,000$ plaques, too many to count.

Assay of dilution C will give: $12 * 10^6 * 0.1 * (1/1000) = 1200$ plaques, too many to count.

Assay of dilution D will give: $12 * 10^6 * 0.1 * (1/10\,000) = 120$ plaques, so this is the dilution which should be used.

(d) Work out a simpler way of obtaining the same final dilution.

Dilution A: 1 mL + 9 mL   1:10, total dilution 1:10

Dilution B: 1 mL + 9 mL   1:10, total dilution 1:100

Dilution C: 1 mL + 9 mL   1:10, total dilution 1:1000

Dilution D: 1 mL + 9 mL   1:10, total dilution 1:10 000.

# Chapter 5

**5.1.** A snail travels in an elliptical path where the longest diameter is 1 m and the shortest diameter 0.5 m. How far does the snail travel?

Length of an elipse:

$$\pi[1.5(x + y) - \sqrt{(x * y)}]$$
$$\pi[1.5(1 + 0.5) - \sqrt{(1 * 0.5)}] = 4.85\,\text{m}$$

**5.2.** The wing cases of a new species of beetle are shaped like right-angled triangles with sides of 1, 1.5 and 1.75 cm. What is the area of the wing cases? (Remember a beetle has two wing cases.)

Area of a triangle:

$$0.5(x * y)$$
$$0.5(1 * 1.5) * 2 = 1.5\,\text{cm}^2$$

**5.3.** An aquarium has internal dimensions of $109 * 47 * 47$ cm. What is its volume?

Volume of a cuboid:

$$x * y * z$$
$$109 * 47 * 47 = 240\,781\,\text{cm}^3 = 240.78\,\text{L}$$

**5.4.** A conical antheap has a base area of $0.65\,\text{m}^2$ and a height of 0.24 m. What volume does the antheap occupy?

Volume of a cone:

$$b * h/3$$
$$(0.65 * 0.24)/3 = 0.052\,\text{m}^3$$

**5.5.** An onion root tip has a circumference of 19.9 μm at its base and a length of 9.2 μm (the tip has not been cut off from the plant, so you may assume it is a baseless cone). What are the surface area and volume of the root tip?

(i) Surface area of a cone:

$$0.5(p * h)$$
$$0.5(19.9 * 9.2) = 91.54\,\mu m^2$$

(ii) Circumference at base $= 2\pi r = 19.9\,\mu m^2$

Therefore $r = 19.9/2\pi = 3.17\,\mu m$.
Area of base $= \pi r^2 = 31.57\,\mu m^2$.
Volume $= b * h/3 = (31.57 * 9.2)/3 = 96.8\,\mu m^3$.

**5.6.** Lymphocytes (white blood cells) are essentially spherical in shape. In a blood sample, there are $2 * 10^6$ lymphocytes mL$^{-1}$. If the average diameter of a lymphocyte is 7.5 μm, what is the total surface area and total volume of all the lymphocytes in 1 mL of blood?

(a) Surface area of a cell:

$$4\pi r^2$$
$$4 * \pi * 3.75^2 = 176.74 * 2 * 10^6 = 353\,480\,000\,\mu m^2$$

(b) Volume of a cell:

$$4\pi r^3/3$$
$$(4 * \pi * 3.75^3)/3 = 220.92 * 2 * 10^6 = 441\,840\,000\,\mu m^3$$

(c) Surface area in cm$^2$:

$$1\,cm^2 = 1 * 10^8\,\mu m^2 \quad 353\,480\,000\,\mu m^2 = 3.53\,cm^2$$

(d) Volume in cm$^3$:

$$1\,cm^3 = 1 * 10^{12}\,\mu m^3 \text{ so: } 441\,840\,000\,\mu m^3 = 0.00\,044\,cm^3$$

**5.7.** A waterlily pad is 10.5 cm in diameter and has an average thickness of 1.5 mm. What volume does it occupy and what is its total surface area? (You may assume the lily pad is a complete circle.)

Standardize the units – in this case, centimetres:
Volume of a cylinder:

$$\pi r^2 h$$
$$\pi(5.25^2 * 0.15) = 12.99\,cm^3$$

Surface area of a cylinder:

$$2\pi rh + 2\pi r^2$$
$$\{2[\pi(5.25 * 0.15)]\} + 2\pi 5.25^2 = 4.95 + 173.18 = 178.13\,\text{cm}^2$$

5.8. A $20\,\mu\text{L}$ blood sample from a mouse contains $9 * 10^4$ erythrocytes (red blood cells). If the mouse contains a total of $2.5\,\text{mL}$ of blood and the average volume of an erythrocyte is $9 * 10^{-11}\,\text{L}$, calculate:

(a) The number of erythrocytes per mL of blood.

$$1\,\text{mL} = 1000\,\mu\text{L}/20 = 50$$

$$9 * 10^4 * 50 = 4.5 * 10^6$$

(b) The total number of erythrocytes in the mouse:

$$4.5 * 10^6 * 2.5 = 1.125 * 10^7$$

(c) The total volume of the erythrocytes in the mouse:

$$1.125 * 10^7 * 9 * 10^{-11} = 0.00101\,\text{L} = 1.01\,\text{mL}$$

5.9. The Earth is, on average, about $1.5 * 10^8\,\text{km}$ from the Sun. Earth's orbit is actually elliptical, and has different speeds at different parts of the orbit, but for this calculation, assume that the earth has a circular orbit.

(a) How many metres does the earth travel in one year?

Length of orbit:

$$2\pi r \ (\text{or } \pi * d)$$
$$2 * \pi * 1.5 * 10^8\,\text{km} = 9.42 * 10^8\,\text{km} \ (= 9.42 * 10^{11}\,\text{m})$$

(b) How far does the earth travel in one month?

$$9.42 * 10^8\,\text{km}/12 = 7.85 * 10^7\,\text{km} \ (= 9.42 * 10^{11}\,\text{m}/12 = 7.85 * 10^{10}\,\text{m})$$

5.10. The Earth rotates on its axis once every $24\,\text{h}$. A person at the equator travels through space in a (nearly) circular path with a radius of $6400\,\text{km}$. What is the person's speed?

Distance travelled:

$$2\pi r \ (\text{or } \pi * d)$$
$$2 * \pi * 6400\,\text{km} = 40\,212\,\text{km}$$

Speed:

$$40\,212\,\text{km}/24\,\text{h} = 1675\,\text{km}\,\text{h}^{-1}$$
$$= (1675\,\text{km} * 1000\,\text{m})/3600\,\text{s} = 465\,\text{m}\,\text{s}^{-1}$$

# Chapter 6

**6.1.** Simplify (calculate the value of): $8^4 * 8^4$

$$8^4 * 8^4 = 8^{4*4} = 8^8 \text{ (add the exponents).}$$

**6.2.** Simplify: $8^5/8^4$

$$8^5/8^4 = 8^{5-4} = 8^1 \text{ (subtract the exponents).}$$

**6.3.** Simplify: $(8^5)^5$

$$(8^5)^5 = 8^{25} \text{ (multiply the exponents).}$$

**6.4.** Simplify: $\log_{10}(5 * 4)$

$$\log_{10}(5 * 4) = \log_{10} 5 + \log_{10} 4 = 0.699 + 0.602 = 1.301 \text{ (add the logs).}$$

**6.5.** Simplify: $\log_{10}(5/4)$

$$\log_{10}(5/4) = \log_{10} 5 + \log_{10} 4 = 0.699 - 0.602 = 0.097 \text{ (subtract the logs).}$$

**6.6.** Simplify: $\log_{10}(3.3)^3$

$$\log_a(x)^n = n \log_a(x)$$
$$3 \log_{10}(3.3) = 3 * 0.519 = 1.56$$

**6.7.** What is the pH of a 0.011 M solution of HCl?

$$pH = -\log[H^+]$$
$$-\log 0.011 = -(-1.959) = 1.959$$

**6.8.** What is the pH of 100 mL of a solution containing 9 mg of HCl? The molecular weight of HCl is 36.46.

(a) Molarity of HCl $= 0.009/36.46 * 10 = 0.0025$ M.

(b) pH $= -\log 0.0025 = -(-2.61) = 2.61$.

**6.9.** What is the $H^+$ concentration in a solution of HCl with a pH of 3?

$$pH = -\log[H^+]$$

So concentration $= $ antilog $- [H^+] = -$ antilog $- 3 = 0.001$ M.

**6.10.** In an exponentially growing bacterial culture:

Number of cells $mL^{-1}$ ($N_0$) at 3 p.m. ($t_0$) $= 5.5 * 10^3$

Number of cells $mL^{-1}$ ($N$) at 5 p.m. ($t$) $= 2.5 * 10^6$

Calculate:

(a) $\log_{10} N_0$

$$\log_{10} 5.5 * 10^3 = 3.74$$

(b) $\log_{10} N$

$$\log_{10} 2.5 * 10^6 = 6.40$$

(c) $t - t_0$

$$5 - 3 = 2(h)$$

(d) $\log_{10} N - \log_{10} N_0$

$$6.40 - 3.74 = 2.66$$

(e) $\mu$

$$\mu = [(\log_{10} N - \log_{10} N_0)\ 2.303]/(t - t_0)$$
$$\mu = (2.66 * 2.303)/2 = 6.126/2 = 3.06\,h^{-1}$$

**6.11.** A skeleton recovered from a peat bog is examined by police forensic scientists to determine whether a crime has been committed. Based on the decay of $^{14}C$ (half-life 5730 years), they estimate that 49% of the original $^{14}C$ in the skeleton has decayed. Calculate the probable age of the skeleton.

$$N = N_0 e^{-\lambda t}$$

For $^{14}C$:

$$e^{\lambda(5730)} = 0.5$$
$$\ln\left(e^{\lambda(5730)}\right) = \ln 0.5$$
$$(5730)\lambda = \ln 0.5$$
$$\lambda = \ln 0.5/5730 = -0.000121$$

Since 49% of the $^{14}C$ has decayed, 51% remains, so:

$$e^{\lambda t} = 0.51$$
$$\ln\left(e^{\lambda t}\right) = \ln 0.51$$
$$\lambda t = \ln 0.51$$
$$t = \ln 0.51/-0.000121 = 5566 \text{ years}$$

**6.12.** The population of a wild mink on a river system increases from 2300 to 3245 in 7 years.

(a) What was the population at the end of the first year?

$$N(t) = N_0\, e^{\lambda t}$$

$N_0 = 2300$, $N = 3245$, $t = 7$ years, so:

$$3245 = 2300\, e^{\lambda t}$$
$$3245/2300 = e^{\lambda t}$$
$$\ln(3245/2300) = \ln(e^{\lambda t})$$
$$\ln(3245/2300) = \lambda * 7$$
$$\lambda = [\ln 3245/2300)]/7 = 0.049$$

After one year:

$$2300\, e^{0.049*1} = 2415.$$

(b) How long will it take for the original population to double?

$$4600 = 2300\, e^{\lambda t}$$
$$4600/2300 = e^{\lambda t}$$
$$2 = e^{\lambda t}$$
$$\ln 2 = \ln e^{\lambda t}$$
$$\ln 2 = \lambda t$$
$$t = \ln 2/0.049 = 14.2 \text{ years.}$$

# Chapter 7

**7.1.** Identify which type of variable each of the following parameters corresponds to:

(a) Blood type (A, B, AB, O)

Nominal – blood groups are names with no inherent order.

(b) Number of eggs in a nest

Discrete – birds do not lay half an egg.

(c) Temperature – Celcius

Interval – no true zero point.

(d) Temperature – kelvin

   Ratio – unlike Celsius, the kelvin scale does have a true zero point.

(e) Age

   Ratio – equally spaced intervals with a true zero point.

(f) Questionnaire (e.g. terrible, poor, average, good, very good)

   Ordinal – these categories have rank order but no inherent value.

(g) Gender

   Nominal.

(h) Height

   Continuous – the height of a population varies continuously between different individuals.

(i) Apple variety (e.g. Cox, Discovery, etc.)

   Nominal.

(j) Number of blood cells per mL

   Discrete or continuous – depends on how the measurement is done.

**7.2.** Construct a grouped frequency distribution table for the following data showing the cumulative frequencies and cumulative percentage:

16, 33, 27, 82, 99, 14, 17, 74, 57, 83, 43, 27, 69, 82, 24, 25, 9, 2, 37, 85

| Grouped scores | Frequency ($f$) | Cumulative ($f$) | Cumulative percentage |
|---|---|---|---|
| 91–100 | 1 | 20 | 100 |
| 81–90 | 4 | 19 | 95 |
| 71–80 | 1 | 15 | 75 |
| 61–70 | 1 | 14 | 70 |
| 51–60 | 1 | 13 | 65 |
| 41–50 | 1 | 12 | 60 |
| 31–40 | 2 | 11 | 55 |
| 21–30 | 4 | 9 | 45 |
| 11–20 | 3 | 5 | 25 |
| 1–10 | 2 | 2 | 10 |
| $n$ | 20 | | |

**7.3.** Are the following frequency distribution tables constructed correctly or wrongly?

(a) Correct. For nominal variables, the order in which the categories are listed makes no difference, but it makes sense to list them in order of increasing or decreasing frequency (although listing the groups by name would also be acceptable):

| Group | Frequency |
|---|---|
| Group 1 | 17 |
| Group 2 | 24 |
| Group 3 | 22 |
| Group 4 | 29 |
| Group 5 | 9 |

(b) Wrong. The groups should be continuous throughout the distribution, with no gaps.

| Scores | Frequency | Cumulative frequency |
|---|---|---|
| 1–10 | 6 | 6 |
| 11–20 | 2 | 8 |
| 21–30 | 14 | 22 |
| 41–50 | 27 | 49 |
| 51–60 | 0 | 49 |
| 61–70 | 33 | 82 |
| 71–80 | 17 | 99 |
| 81–90 | 14 | 113 |
| 91–100 | 2 | 115 |

(c) Wrong. Although the groups do not overlap, they are of different sizes.

7.4. For the following dataset:

$$16, 33, 65, 82, 99, 14, 17, 74, 57, 83, 43, 27, 69, 82, 24, 25, 9, 2, 37, 85, 1, 13, 96$$

(a) Calculate the scores at the first ($Q_1$) and third ($Q_3$) quartiles:

$$\text{Percentile rank} = (n+1)P/100$$
$$P_{25}(Q_1) = (23+1) * 25/100 = 6$$
$$P_{75}(Q_3) = (23+1) * 75/100 = 18$$

(b) Calculate the scores at the 45th ($P_{45}$) and 95th ($P_{95}$) percentiles:

$$P_{45} = (23+1) * 45/100 = 11$$
$$P_{95} = (23+1) * 95/100 = 23$$

| Score | Cumulative frequency | Cumulative percentage | |
|---|---|---|---|
| 1 | 1 | 4 | |
| 2 | 2 | 9 | |
| 9 | 3 | 13 | |
| 13 | 4 | 17 | |
| 14 | 5 | 22 | |
| 16 | 6 | 26 | $P_{25}$ ($Q_1$) |
| 17 | 7 | 30 | |
| 24 | 8 | 35 | |
| 25 | 9 | 39 | |
| 27 | 10 | 43 | |
| 33 | 11 | 48 | $P_{45}$ |
| 37 | 12 | 52 | |
| 43 | 13 | 57 | |
| 57 | 14 | 61 | |
| 65 | 15 | 65 | |
| 69 | 16 | 70 | |
| 74 | 17 | 74 | |
| 82 | 18 | 78 | $P_{75}$ ($Q_3$) |
| 82 | 19 | 83 | |
| 83 | 20 | 87 | |
| 85 | 21 | 91 | |
| 96 | 22 | 96 | |
| 99 | 23 | 100 | $P_{95}$ |
| $n$ | 23 | | |

7.5. Which of the following graph types:

scatter plot, histogram, pie diagram

would be suitable for the following datasets?

(a) proportion of university students from different schools: *histogram or pie diagram*

(b) enzyme activity at a range of pH measurements: *scatter plot*

(c) number of undergraduates, 1999–2005: *histogram*

(d) blood pressure against time: *scatter plot*

(e) grouped frequency distribution: *histogram.*

# Chapter 8

Table 8.3 below contains a set of data on the microbiological quality of bottled drinking water. In this study, the number of bacterial colony-forming units per millilitre of bottled water was measured for 120 different water samples:

**8.1.** Construct a grouped frequency distribution table for this dataset:

| CFU mL$^{-1}$ | $f$ | Cumulative $f$ | Cumulative percentage |
|---|---|---|---|
| 0–999 | 3 | 3 | 2 |
| 1000–1999 | 1 | 4 | 3 |
| 2000–2999 | 1 | 5 | 4 |
| 3000–3999 | 1 | 6 | 5 |
| 4000–4999 | 5 | 11 | 9 |
| 5000–5999 | 7 | 18 | 15 |
| 6000–6999 | 16 | 34 | 28 |
| 7000–7999 | 30 | 64 | 53 |
| 8000–8999 | 44 | 108 | 90 |
| 9000–9999 | 12 | 120 | 100 |
| Total | 120 | | |

**8.2.** Plot a frequency distribution histogram of the data:

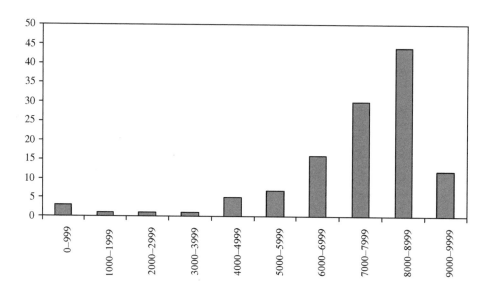

**8.3.** How would you describe this dataset (normal, negative skew or positive skew)?

This dataset is non-symmetrical and has a strong positive skew (right skew), indicating that in this population there is a high frequency of samples containing a high bacterial count.

**8.4.** Calculate (a) the 95th and (b) the 25th percentiles for this dataset:

$$\text{Percentile} = \text{LCB} + \{[(k * n/100) - \text{CFB}]/f\} * I \quad \text{(Chapter 7)}$$

From the table (answer 8.1):

(a) For $P_{95}$ $k * n/100 = 95 * 120/100 = 114$

$$P_{95} = 9000 + [(114 - 108)/12] * 1000 = 9500$$

(b) For $P_{25}$ $k * n/100 = 25 * 120/100 = 30$

$$P_{25} = 6000 + [(30 - 18)/16] * 1000 = 6750$$

**8.5.** Calculate (a) the mean, (b) the median and (c) the mode for this dataset:

(a) Mean $= \mu_x = (\sum X)/N = 889\,866/120 = 7416$.

(b) Median $= P_{50}$ $(Q_2)$ $k * n/100 = 50 * 120/100 = 60$, so:

$$7000 + [(60 - 34)/30] * 1000 = 7867$$

(c) Mode $= 8259$ (two occurrences).

**8.6.** Calculate (a) the range, (b) the semi-interquartile range, (c) the variance and (d) the standard deviation for this dataset:

(a) Range: $9726 - 383 = 9343$.

(b) Semi-interquartile range $(Q_2 - Q_1)$: $7867$ (answer 8.5b) $- 6750$ (answer 8.4b) $= 1117$.

(c) Variance: $\sigma^2 = \sum(X - \mu_x)^2/N = 3\,260\,002$.

(d) Standard deviation: $\sigma_x = \sqrt{\sum(X - \mu_x)^2/N} = \sqrt{3\,260\,002} = 1806$.

**8.7.** Exploratory data analysis:

(a) Construct a scatter plot of the following dataset. Are the data normally distributed?

| $x$ | 10 | 8 | 13 | 9 | 11 | 14 | 6 | 4 | 12 | 7 | 5 |
|---|---|---|---|---|---|---|---|---|---|---|---|
| $y$ | 11.2 | 8.1 | 9.7 | 9.8 | 12.8 | 8 | 6.2 | 3 | 11.8 | 7.3 | 5.7 |

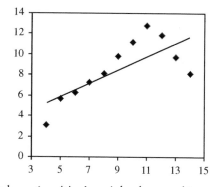

This data has a 'positive' or right skew and is not normally distributed.

(b) Construct a frequency histogram of the following dataset. Are the data normally distributed?

| x | 1–10 | 11–20 | 21–30 | 31–40 | 41–50 | 51–60 | 61–70 | 71–80 | 81–90 | 91–100 |
|---|------|-------|-------|-------|-------|-------|-------|-------|-------|--------|
| y | 0 | 0 | 1 | 4 | 6 | 9 | 13 | 9 | 8 | 4 |

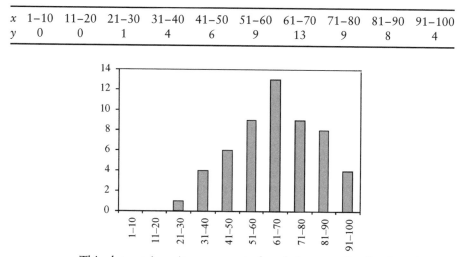

This dataset is quite symmetrical and shows a distribution close to normal.

(c) Construct a stem and leaf diagram of the following dataset. Are the data normally distributed?

21, 23, 25, 26, 26, 27, 29, 31, 32, 32, 33, 35, 35, 36, 37, 38, 38, 39, 41, 41, 41, 42, 42, 44, 45, 47, 48, 48, 49, 51, 52, 53, 53, 53, 54, 55, 55, 55, 57, 61, 62, 63, 66, 71, 74, 91.

```
2|1 3 5 6 6 7 9
3|1 2 2 3 5 5 6 7 8 8 9
4|1 1 1 2 2 4 5 7 8 8 9
5|1 2 3 3 3 4 5 5 5 7
6|1 2 3 6
7|1 4
8|
9|1
```

This dataset has a 'negative' or left skew and is not normally distributed.

(d) Sketch a box and whisker plot of the following dataset. Are the data normally distributed?

14, 20, 22, 25, 27, 28, 31, 33, 38, 42, 51, 53, 61, 62, 65, 71, 74, 77, 78, 84, 86, 91.
Median = 52, first quartile = 29, third quartile = 73.

The box and whisker plot looks like this:

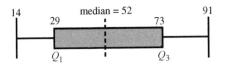

From this, the data looks to be normally distributed. However, if you plot a frequency distribution histogram of this data:

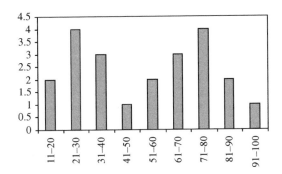

It is clear that the data has a bimodal distribution. This example has been included to illustrate how careful you need to be when assessing data prior to analysis.

# Chapter 9

**9.1.** Cystic fibrosis (CF) is the most common recessive genetic disorder in Caucasians – approximately one person in 2500 carries one copy of the CF gene, which occurs with equal frequency in males and females. If a couple are both carriers of the CF gene and have a child, the following probabilities apply: normal child, non-carrier $P = 0.25$; normal child, carrier, $P = 0.50$; child with cystic fibrosis, $P = 0.25$. What is the probability that the couple will have:

(a) Two children (either sex) who do not carry the CF gene?

$P(\text{non-carrier})$ and $P(\text{non-carrier}) = 0.25 * 0.25 = 0.0625$

(b) One son who is a carrier?

$$P(\text{boy}) \text{ and } P(\text{carrier}) = 0.5 * 0.5 = 0.25$$

(c) Two daughters, one who is a carrier and one who has cystic fibrosis?

$$P(\text{girl and carrier}) \text{ and } P(\text{girl and CF})$$
$$= (0.5 * 0.25) * (0.5 * 0.25)$$
$$= (0.25 * 0.125)$$
$$= 0.03125$$

(d) Two daughters with cystic fibrosis?

$$P(\text{girl and CF}) \text{ and } P(\text{girl and CF})$$
$$= (0.5 * 0.25) * (0.5 * 0.25)$$
$$= (0.125 * 0.125)$$
$$= 0.015625$$

These probabilities are controlled by the 'product/and' rule: the probability of several distinct events occurring successively or jointly is the product of their individual probabilities, provided that the events are independent.

9.2. In order to study great crested newt (*Triturus cristatus*) populations, 150 newts are harmlessly marked with a temporary non-toxic dye. Fifteen newts are then returned to each of 10 ponds known to contain this species. One week later, the ponds are fished again and, of 351 newts caught, 54 are marked.

(a) Estimate the total population of great crested newts in these 10 ponds.

This problem can be solved using proportional frequency: $54/150 = 0.36$, i.e. 36% of the marked newts were recovered.

$$\text{Total population} = 351 * (1/0.36) = 975$$

(b) If one pond has a population of 107 newts (15 marked), what is the probability of catching marked (M) and unmarked (U) newts in this order: UUMUUUMU?

$$P(\text{M}) = 15/107 = 0.14$$
$$P(\text{U}) = (107 - 15)/107 = 0.86$$
$$[\text{N.B. } P(\text{U}) \text{ also} = 1 - P(\text{M}) = 1 - 0.14 = 0.86]$$

Since this involves selection without replacement, use the AND rule:

$$P(\text{UUMUUUMU}) = 0.86 * 0.86 * 0.14 * 0.86 * 0.86 * 0.86 * 0.14 * 0.86$$
$$= 0.0079$$

**9.3.** In a health survey, 19 of 60 men and 12 of 40 women are found to smoke cigarettes.

(a) What is the probability of a randomly selected individual being a male who smokes?

This is a joint probability. The number of male smokers divided by the total $= 19/100 = 0.19$.

(b) What is the probability of a randomly selected individual smoking?

This is the total number of smokers divided by the total $= 31/100 = 0.31$.

(c) What is the probability of a randomly selected male smoking?

This is the number of male smokers divided by the number of males $= 19/60 = 0.317$.

(d) What is the probability that a randomly selected smoker is male?

There are 19 males out of 31 smokers, so $19/31 = 0.613$.

**9.4.** The probability of being infected with HIV from each single exposure to one of the following events is approximately: unprotected sexual intercourse with an HIV carrier – 0.005; sharing an infected needle for intravenous drug use – 0.007; needlestick injuries in healthcare workers – 0.003. The cumulative probability of being infected $P(i)$ after $n$ occurrences is given by the formula:

$$P(i) = 1 - (1 - k)^n$$

where $k =$ the probability of being infected with HIV from each single exposure and $n =$ the number of occurrences. What is the probability of being infected with HIV after:

(a) Five occurrences of unprotected sexual intercourse with an HIV carrier:

$$P(i) = 1 - (1 - 0.005)^5$$
$$= 1 - (0.995)^5$$
$$= 1 - 0.975$$
$$= 0.025$$

(b) Nine occurrences of sharing an infected needle for intravenous drug use:

$$P(i) = 1 - (1 - 0.007)^9$$
$$= 1 - (0.993)^9$$
$$= 1 - 0.939$$
$$= 0.061$$

(c) One needlestick injury in a healthcare worker who subsequently has unprotected sexual intercourse with an HIV carrier three times:

Here the probabilities are additive

$$P_{needle} = 0.003$$
$$P_{sex} = 1 - (1 - 0.005)^3$$
$$= 1 - (0.995)_3$$
$$= 1 - 0.985$$
$$= 0.015$$

Overall probability of infection

$$0.003 + 0.015 = 0.018$$

**9.5.** In a practical class, you are given three tubes of an enzyme (A B C) needed to perform an experiment you only have time to do once. A kind demonstrator has told you that only one of the tubes contains active enzyme and the other two are inactive. You choose tube A. To help you further, the demonstrator tells you that tube B contains inactive enzyme. Should you stick with tube A or switch to tube C for the experiment? (Explain why.)

Answer: switch to tube C. Why?

Originally, the probability of picking the tube containing active enzyme was $1/3 = 0.033$. The demonstrator has told you that tube B is inactive. If you do not switch tubes, the probability remains at 0.033. However, since one inactive tube has been eliminated, if you switch tubes, the probability of using the active enzyme is now $1/2 = 0.5$. This is a version of the famous 'Monty Hall problem' named after an American quiz show host. If you don't think this is the right answer don't worry – lots of people have difficulty with this. This simple problem is one of the most conceptually difficult in probability theory and the answer seems to be counterintuitive. Find a friend and try the experiment for yourself.

# Chapter 10

**10.1.** The heights of a group of girls and a group of boys was measured. The frequency of measurements in both groups was found to have a normal distribution:

|                    | Girls   | Boys   |
|--------------------|---------|--------|
| Mean               | 1.25 m  | 1.29 m |
| Standard deviation | 6 cm    | 5 cm   |

(a) Susan's height is 1.31 m. What is her $z$-score?

$$z = (\text{score} - \text{mean})/\text{standard deviation}$$
$$= (1.31 - 1.25)/0.06 = 1$$

(b) Michael's height is 1.31 m. What is his $z$-score?

$$(1.31 - 1.29)/0.05 = 0.4$$

(c) Sally's $z$-score is $-1.2$. Is she taller or shorter than the average for her group?

Since her $z$-score is negative, she must be shorter than the average.

(d) True or false: the boys' $z$-scores are higher than the girls' $z$-scores (explain your answer).

False. The $z$-scores for both groups are they same since they are standardized by the mean and standard deviation of a normal population.

(e) What percentage of boys are taller than 1.39 m?

$$z\text{-score} = (1.39 - 1.29)/0.05 = 2 \text{ standard deviations}$$

Since 95% of a normally distributed population is covered by $\pm 2$ standard deviations (chapter 8), 2.5% of boys will be taller than 1.39 m ($z$-score $= 2$).

**10.2.** A group of 12 patients with high blood pressure is treated with drug A for 3 months. At the end of the treatment period, their blood pressure is measured and treatment with drug B started. After a further 3 months, their blood pressure is measured again. Analyse the data from this trial using Student's $t$-test:

|            | Drug A | Drug B |
|------------|--------|--------|
| Patient 1  | 189    | 186    |
| Patient 2  | 181    | 181    |
| Patient 3  | 175    | 179    |
| Patient 4  | 186    | 189    |
| Patient 5  | 179    | 175    |
| Patient 6  | 191    | 189    |
| Patient 7  | 180    | 183    |
| Patient 8  | 183    | 181    |
| Patient 9  | 183    | 186    |
| Patient 10 | 189    | 190    |
| Patient 11 | 176    | 176    |
| Patient 12 | 186    | 183    |

(a) What sort of $t$-test should you perform to analyse these data?

Since the two sets of measurements refer to the same patients, there is a direct one-to-one correspondence between the groups and a paired $t$-test should be used.

(b) Should you use a one tailed or two-tailed test?

Since we have no means of predicting the outcome of the trial, a two-tailed test should be used.

(c) How many degrees of freedom are there in this test?

$$\text{df} = n - 1 = 12 - 1 = 11$$

(d) Is there a statistically significant difference at the 95% confidence level in the blood pressure of the patients after treatment with the two drugs?

Before starting the test, are the data normally distributed?

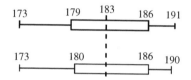

Yes, approximately.

$$t = \frac{d_{av}}{SD/\sqrt{N}}$$

$$t_{calc} = 0/(2.8/3.46) = 0/0.81 = 0$$

$$t_{crit} = 2.2 \ (\text{Appendix 3})$$

Since the calculated value of $t$ ($t_{calc}$) is less than the critical value of $t$ ($t_{crit}$) for this situation, there is no evidence of a statistically significant difference between the two drugs.

10.3. In a study of the acidification of lakes, pH measurements were made of a series of lakes draining into two different rivers, A and B. Analyse the data from this trial using Student's $t$-test:

| A | | B | |
|---|---|---|---|
| 6.97 | 7.2 | 5.93 | 6.70 |
| 5.88 | 7.81 | 4.88 | 6.81 |
| 6.41 | 6.98 | 5.71 | 6.18 |
| 6.85 | 7.42 | 5.85 | 6.42 |
| 6.24 | 5.59 | 5.24 | 4.59 |

| A | | B | |
|---|---|---|---|
| 6.26 | 6.77 | 7.86 | 6.77 |
| 5.01 | 5.84 | 4.01 | 5.24 |
| 7.64 | 8.41 | 6.64 | 7.31 |
| 6.40 | 6.59 | 7.20 | 6.29 |
| 6.72 | 7.10 | 6.32 | 6.10 |

(a) What sort of $t$-test should you perform to analyse these data?

Since there is no direct correspondence between the groups, an unpaired $t$-test should be used.

(b) Should you use a one-tailed or two-tailed test?

Since we have no means of predicting the outcome of the trial, a two-tailed test should be used.

(c) How many degrees of freedom are there in this test?

$$df = (nA + nB) - 2 = (20 + 20) - 2 = 38$$

(d) Is there a statistically significant difference at the 95% confidence level in the pH readings of the lakes draining into the two rivers?

Before starting the test, are the data normally distributed?

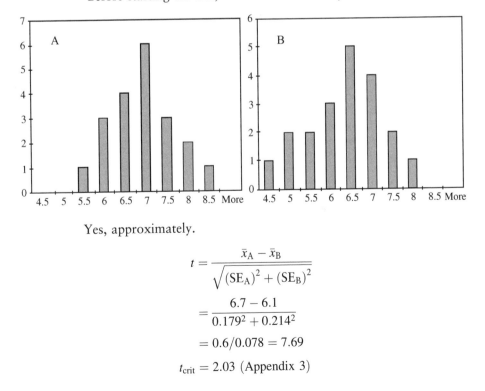

Yes, approximately.

$$t = \frac{\bar{x}_A - \bar{x}_B}{\sqrt{(SE_A)^2 + (SE_B)^2}}$$

$$= \frac{6.7 - 6.1}{0.179^2 + 0.214^2}$$

$$= 0.6/0.078 = 7.69$$

$$t_{crit} = 2.03 \text{ (Appendix 3)}$$

Since the calculated value of $t$ ($t_{calc}$) is greater than the critical value of $t$ ($t_{crit}$) for this test, there is evidence of a statistically significant difference between in the pH readings of the lakes draining into the two rivers.

**10.4.** The number of eggs in robins' nests in three different areas of woodland were counted and found to be:

A: 2, 0, 1, 1, 1, 3, 1, 3, 2, 1, 1, 2, 2, 2, 1, 3, 3, 1, 2, 0, 1, 1, 1, 1, 0
B: 2, 1, 2, 0, 1, 5, 1, 2, 3, 2, 1, 2, 2, 2, 0, 3, 2, 0, 1, 1, 0, 1, 0, 0, 1
C: 2, 0, 2, 0, 2, 5, 1, 2, 2, 1, 0, 1, 3, 2, 3, 2, 1, 1, 0, 1, 2, 1, 1, 4, 2

Can you perform an ANOVA test to demonstrate whether or not there a statistically significant difference at the 95% confidence level between the three woodlands?

Since ANOVA is a parametric test, before starting we need to ensure the data is normally distributed:

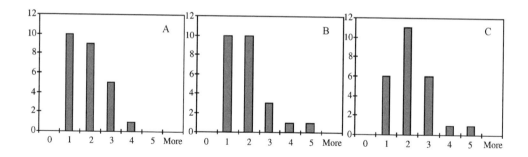

Each of these datasets has a strong positive skew. If we were to perform an ANOVA test on this data, the results would be meaningless (unless the data were transformed to a normal distribution first).

**10.5.** A biologist measures the preference of three-spined sticklebacks (*Gasterosteus aculeatus*) for various food items. In a 3 h period, fish of length less than 4 cm consumed 14 *Daphnia galeata*, 14 *Daphnia magna* and 36 *Daphnia pulex*, while fish longer than 4 cm consumed 6 *Daphnia galeata*, 24 *Daphnia magna* and 31 *Daphnia pulex*. Use the $\chi^2$-test to compare the distribution of these variables and decide whether there is a statistically significant difference at the 95% confidence level between the feeding behaviour of the larger and the smaller sticklebacks.

(a) Construct a contingency table for the data.

| Fish size | *Daphnia galeata* | *Daphnia magna* | *Daphnia pulex* | Total |
|---|---|---|---|---|
| < 4 cm | 14 | 14 | 36 | 64 |
| > 4 cm | 6 | 24 | 31 | 61 |
| Total | 20 | 38 | 67 | 125 |

(b) Formulate the null hypothesis for this experiment.

There is no significant difference between the feeding behaviour of larger and smaller sticklebacks.

(c) How many degrees of freedom are there in this case?

$$df = (\text{number of columns} - 1) * (\text{number of rows} - 1)$$
$$= (3 - 1) * (2 - 1) = 2$$

(d) Calculate $\chi^2$.

Expected frequencies: (column total $*$ row total)/overall total

| Fish size | *Daphnia galeata* | *Daphnia magna* | *Daphnia pulex* | Total |
|---|---|---|---|---|
| < 4 cm | 10 | 19 | 34 | 64 |
| > 4 cm | 10 | 19 | 33 | 61 |
| Total | 20 | 38 | 67 · | 125 |

$\chi^2$ calculations: (actual $-$ expected)$^2$/expected

| Fish size | *Daphnia galeata* | *Daphnia magna* | *Daphnia pulex* |
|---|---|---|---|
| < 4 cm | 1.381 | 1.530 | 0.084 |
| > 4 cm | 1.449 | 1.605 | 0.088 |

$$\chi^2 = \sum \frac{(\text{observed frequency} - \text{expected frequency})^2}{\text{expected frequency}}$$
$$= 1.381 + 1.53 + 0.084 + 1.449 + 1.605 + 0.088 = 6.14$$

(e) Is there a statistically significant difference at the 95% confidence level between the feeding behaviour of the larger and the smaller sticklebacks?

Critical value of $\chi^2$(Appendix 3) $= 5.99$

Since the calculated value of $\chi^2$ (6.14) is greater than the critical value (5.99), the null hypothesis is rejected, i.e. there is a statistically significant difference at the 95% confidence level between the feeding behaviour of the larger and the smaller sticklebacks.

**10.6.** A group of 353 cancer patients are treated with a new drug. of the patients who receive this treatment, 229 survive for more than 5 years after the commencement of treatment. Compare this result with a control group of 529 similar patients treated with the previously accepted drug therapy, 310 of whom survive for more than 5 years after the commencement of treatment. Is there a statistically significant difference at the 95% confidence level between the survival rates of the patients who received the new drug and those who received the previously accepted therapy?

Construct a contingency table for the data. Observed frequencies:

| Survival | New drug | Old drug | Total |
|----------|----------|----------|-------|
| > 5 years | 229 | 124 | 353 |
| < 5 years | 310 | 219 | 529 |
| Total | 539 | 343 | 882 |

Formulate the null hypothesis: there is no significant difference in survival rate between the two drug treatments.

Degrees of freedom:

$$df = (\text{number of columns} - 1) * (\text{number of rows} - 1)$$
$$= (2 - 1) * (2 - 1) = 1$$

Calculate $\chi^2$: expected frequencies (column total * row total)/overall total

| Survival | New drug | Old drug | Total |
|----------|----------|----------|-------|
| > 5 years | 216 | 137 | 353 |
| < 5 years | 323 | 206 | 529 |
| Total | 539 | 343 | 882 |

$\chi^2$ calculations: $(\text{actual} - \text{expected})^2/\text{expected}$

| Survival | New drug | Old drug |
|----------|----------|----------|
| > 5 years | 0.82 | 1.28 |
| < 5 years | 0.55 | 0.86 |

$$\chi^2 = 0.82 + 1.28 + 0.55 + 0.86 = 3.51$$

Critical value of $\chi^2$ (Appendix 3) = 3.84. Since the calculated value of $\chi^2$ (3.51) is less than the critical value (3.84), the null hypothesis is accepted, i.e. there is no statistically significant difference at the 95% confidence level in survival rate between the two drug treatments.

# Chapter 11

**11.1.** Construct scatter plots of the four datasets in Table 11.2 and state whether each is suitable for correlation analysis or not.

(A) This data has an $r = 0.064$, a weak positive correlation, and is suitable for analysis:

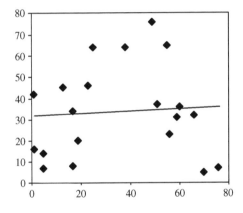

(B) This data has an $r = 0.62$, a moderate positive correlation, and is suitable for analysis:

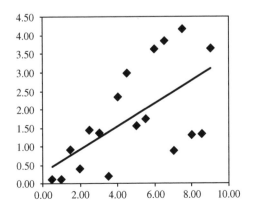

(C) You cannot calculate a correlation coefficient for these data:

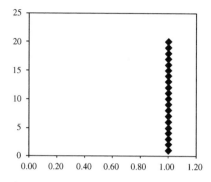

(D) This data has an $r = -0.684$, a strong negative correlation, and is suitable for analysis:

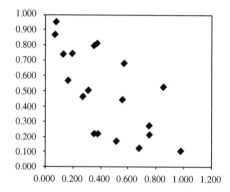

**11.2.** In a biology practical, the lengths and weights of 10 earthworms are measured (Table 11.3). Determine the Pearson correlation coefficient for this data. What can you say about the relationship between the length and the weight of an earthworm?

Start by plotting a scatter graph of the data:

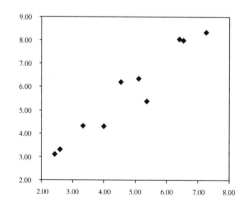

From this, can already see that there appears to be a positive correlation between length and weight.

$$r = \frac{\sum XY - \frac{(\sum x)(\sum y)}{n}}{\sqrt{\left(\sum X^2 - \frac{(\sum x)^2}{n}\right)\left(\sum Y^2 - \frac{(\sum y)^2}{n}\right)}}$$

$$= \frac{301 - [(48 * 57)/10]}{\sqrt{[252 - (2265/10)][364 - (3289/10)]}}$$

$$= 27.4/\sqrt{25.5 * 35.1}$$

$$= 27.4/29.9$$

$$r = 0.916$$

There is a very strong positive correlation between length and weight in these earthworms, and, since the calculated value of $r$ (0.916), is greater than the critical value of $r$ (0.63) for this experiment (df $= n - 2 = 8$, $\alpha = 0.05$; Appendix 3), this is a statistically significant result.

**11.3.** The lengths of the femur and humerus were measured and compared in 12 human skeletons. The data are presented here in rank order only, where 1 = shortest, 12 = longest (Table 11.4). Is there a statistically significant correlation between the length of the two bones?

Since the data are presented in rank order rather than in raw form, the Spearman rank order correlation will be used to examine the relationship between them. Construct a scatter plot:

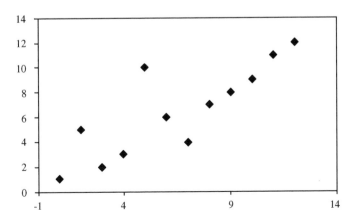

There appears to be a positive correlation between the lengths of the two bones. The Spearman calculation is:

$$r_s = 1 - \frac{6 \sum D^2}{N(N^2 - 1)}$$

| Subject | Length of femur | Length of humerus | Difference (D) | $D^2$ |
|---------|-----------------|-------------------|----------------|-------|
| 1 | 3 | 2 | 1 | 1 |
| 2 | 5 | 10 | − 5 | 25 |
| 3 | 2 | 5 | − 3 | 9 |
| 4 | 11 | 11 | 0 | 0 |
| 5 | 4 | 3 | 1 | 1 |
| 6 | 6 | 6 | 0 | 0 |
| 7 | 10 | 9 | 1 | 1 |
| 8 | 9 | 8 | 1 | 1 |
| 9 | 8 | 7 | 1 | 1 |
| 10 | 1 | 1 | 0 | 0 |
| 11 | 7 | 4 | 3 | 9 |
| 12 | 12 | 12 | 0 | 0 |
| | | | Total | 48 |

$$r_s = 1 - (6 * 48)/12(12^2 - 1)$$

$$= 1 - (288/1716)$$

$$= 1 - 0.168$$

$$r_s = 0.832$$

Critical value of $r$ (df $= 12 - 2 = 10$, $\alpha = 0.05$) $= 0.576$ (Appendix 3), so the null hypothesis, i.e. there is no proof of an association between the lengths of the two bones, is rejected.

11.4. The reaction rate of the enzyme lactate dehydrogenase was measured at different concentrations of the substrate, pyruvate (Table 11.5). Using statistics software (Appendix 2) the $r^2$ value for this experiment was determined as 0.939. How would you describe the accuracy of the results of the experiment?

While an $r^2$ value of 0.939 represents a strong correlation, this is certainly no higher than a skilled biochemist would expect to obtain for such an experiment:

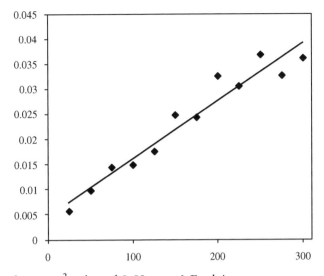

**11.5.** What does an $r^2$-value of 0.59 mean? Explain.

Fifty-nine per cent of the variance in X can be explained by variation in Y. Likewise, 59% of the variance in Y can be explained by variation in X, and 59% of the variance is shared between X and Y. Forty-one per cent of the variation is therefore not explained by the regression model ($100 - 59 = 41$). This result would therefore normally be classed as a medium-strength correlation.

# Appendix 2: Software for Biologists

Computers are now an inescapable part of everyday life, from your medical details to your bank account. Although for a while biology lagged behind other sciences such as physics in the application of computers, this has now changed and 'computational biology' or 'bioinformatics' is now at the forefront of the digital information revolution. What use do computers have in biology? Almost anything to do with the digital storage, classification and analysis of biological data uses computers. In addition, computers are increasingly about communications, through email, mobile phones and digital radio and television.

Although they have become integrated into the fabric of life and work, at the most basic level, computers are merely a tool. Like any other tool, it pays to know how to use them properly. Although many of the general aspects of the computer revolution are beyond the scope of this book, others are highly relevant. However, to begin, we will look at the more general uses of computers then move on to the more specific aspects.

## E-mail

Electronic mail (e-mail) is a powerful and rapid means of sending and receiving information, including not only text files but also complete computer files such as pictures, sounds or videos as 'attachments' to messages. What email software you use is likely to be determined by local circumstances, but many programs are available, and some of the best are free, e.g. Microsoft® Outlook Express® (available from www.microsoft.com).

Email seems easy, but there is more to it than meets the eye:

- DO NOT SHOUT. THAT MEANS, DON'T WRITE IN CAPITALS. Turn the caps lock off. You can use uppercase to stress a point, but not the whole message.

- Do not send formatted email (i.e. HTML format) unless you know the person you are sending it to wants to receive it. Set your email preferences to plain text.

- Remember to fill in the subject line of the message with a relevant subject or at best you will look like a beginner, and at worse your messages will simply be ignored.

- Spelling and grammar – Text messaging is regarded as cool by some, but if you use text messaging-style abbreviations will the person you are writing to be impressed by your coolness? Write in full sentences, use the in-built spellchecker and put a little thought into composing your message. Email lets the recipient build a mental picture of you as a person very quickly.

- Familiarity – email should be fun, but remember who you are talking to. First names are fine for friends and personal messages, but 'professional' messages to your bank manager or tutors should follow the usual formalities. The best guide is to address someone by email in the same way to would talk to them face to face.

- Getting the message – 97% of human communication is said to be non-verbal (smiles, body posture, tone of voice, etc.). With email, you lose all that, and it is easy to be misunderstood. What you intended as a joke can easily be misinterpreted as offensive. A whole range of 'emoticons' (☺) and acronyms have evolved to help with this. Use them to avoid misunderstandings, but when sending formal email, it is best to play on the safe side.

- Recipients – check you are sending the message to the person(s) it is intended for. Make sure you understand the difference between the Reply (reply to only the sender of the message) and Reply to all (reply to all recipients of the To and Cc. boxes of the message). If an email requires acknowledgement, then you should reply promptly, but do not automatically respond to all messages if a reply is not needed. With email, less is more.

- Privacy – there is no such thing as a private email. Everything you write can potentially be seen by many others, from email administrators to anyone who reads your email after it has been forwarded. Always be prepared to stand by everything you write, so read and think before clicking Send, especially if you are feeling angry at the time.

- Flames and spam – to be flamed means that you have sent an email to a person(s) that has caused that person(s) to respond with a verbal attack in electronic form. If you respond in the same tone, you could start a 'flame war'. There are only two smart ways to respond to flames. If you made a mistake or regret your original message, apologize unreservedly. If not, delete the flame and do not reply. Spam is any form of unwanted and unsolicited email, including virus hoaxes. Only fools reply to spam – just delete it. If you receive email with seriously offensive content (e.g. racial, religious, sexual or threatening), forward it to the administrator of your email system ('postmaster') with an explanation and ask for action to be taken.

## Word processors

Although there are many different word processing applications available, with about a 93% share of the global market, Microsoft® Word® is by far the single most popular word processing software package. It is impossible to cover all of the features of MSWord here, since the program contains around $1 * 10^5$ distinct editing, formatting and other commands – just reading the complete list aloud would take several days. However, several features of MSWord are worthy of special note:

- Spellchecker – no spellchecker is perfect, but the MSWord built-in spellchecker is a valuable part of the program. There is no excuse to EVER submit written work containing elementary spelling and typig (*sic*) errors, so use it. The value of MSWord's built-in grammar checker is more debatable. If you want to turn the grammar check off, go to **Tools: Options: Spelling and Grammar.**

- Graphics – there are two basic types of graphics that you can include in MSWord documents: drawing objects and pictures. Drawing objects include AutoShapes®, curves, lines and 'WordArt'. These objects are part of your Word document. Use the Drawing toolbar to change and enhance these objects with colours, patterns, borders and other effects. Pictures are graphics that were created as another file. They include bitmaps, scanned pictures and photographs, and clip art. You can change and enhance pictures using the options on the Picture toolbar and a limited number of options on the Drawing toolbar.

- Macro viruses – always check documents for macros that might contain viruses. A macro virus is a type of computer virus stored in the macros (sequences of MSWord actions which can be recorded and played back to repeat the actions) within a document or template. When you open the document, the virus is activated, transmitted to your computer and stored in your document template. From that point on, every document you save is infected with the virus, and if other users open these infected documents, the virus is transmitted to their computers as well. It is strongly recommend that you turn the macro virus protection on.

## Presentation and graphics

A common complaint is that many essays and reports could be greatly improved by the addition of relevant figures and diagrams. Although limited graphics tools are available in MSWord, Microsoft PowerPoint® is a more sophisticated presentation graphics program. This program is useful for creating presentations which can be projected onscreen or printed as notes, handouts and overheads. With an increasing emphasis on oral presentation skills, it is useful for students to become familiar with the use of such software.

## Internet resources

It used to be said that the internet is a global telephone network without a directory. This is becoming increasingly untrue as ever more sophisticated search engines make the task of finding online information easier. Internet information is accessed using a web browser such as Netscape® or Microsoft Internet Explorer®. Both are available free of charge at www.netscape.com and www.microsoft.com, respectively. Not using any of the information available on the internet is a huge mistake, but of course not all the information online is reliable. Being able to tell the difference has been called 'information literacy', i.e. the ability to: access information efficiently and effectively; evaluate information critically and competently; use information accurately and creatively.

Points to consider are:

1. Authority – who is the author and what are their qualifications? Does the site reveal the author's education, training and background and explain whether they are a trained expert, experienced enthusiast or an uninformed observer?

2. Bias – is the language emotional, inflammatory, profane or confusing? Does the information represent a single opinion or a range of opinions? Is it likely to have been provided to support a particular agenda or bias?

3. Clarity and order – is the information clearly stated? Does the author define important terms? Is the format is clear, logical and easily navigable? Does the information contain spelling, grammatical or typographical errors?

4. Publisher – who is the publisher and what is the purpose of the site? Does the page include the author's name, title and/or position, organization and contact information, such as an email or postal address? Does the address (URL) of the page include a .gov (government), .edu/.ac (educational/academic), .com (commercial), .org (non-profit organization) address?

5. Timeliness – is the information current? When was it posted and/or last updated?

6. Validity – do the facts presented support the conclusions?

7. Verifiability – does the material contain unsubstantiated general-izations? Are sources provided?

It is pointless to list here the addresses of web pages which are likely to be out of date by this time this book is published. What is needed is for internet users to develop an information search strategy. To a large extent, this is the same for the internet as for interrogating any database:

1. Formulate a question.

2. Write down key search terms or phrases.

3. What is the general subject and what are some related subjects?

**Table A1**   Some of the best resources for biologists on the internet

| Site/URL | Description |
|---|---|
| PubMed: www.ncbi.nlm.nih.gov/entrez/ query.fcgi?db=PubMed | The US National Center for Biotechnology Information (NCBI) free gateway to the MedLine database, containing over 10 000 000 biomedical journal citations |
| ISI Web of Science: www.isinet.com (UK: wos.mimas.ac.uk) | Bibliographic database providing access to over 8500 journals |
| NCBI Entrez Genomes: www.ncbi.nlm.nih.gov/entrez/ query.fcgi?db=Genome | Contains the entire genome sequences and taxonomic information on over 800 organisms – bacteria, archaea, eukaryota, viruses and cell organelles |
| Tree of Life: http://tolweb.org/ tree/phylogeny.html | A collaborative web project containing taxonomic information on all living (and extinct) organisms |

4. How far back in time do you need to look for published information?

5. What scholarly level (level of difficulty) is required?

6. What types of resources are needed (periodicals, books, references, maps, charts, graphics, statistics)?

7. How much information do you need for what you am doing?

8. What databases are appropriate?

Although there are many internet search engines available, without doubt the best is Google™ (www.google.com). Google is not only fast, reliable and huge (over three billion documents indexed), but uses a unique 'PageRank™' system which delivers unbiased information. If you are unfamiliar with the internet, you may wish to register for one of the free online courses available. Among the best of these are Netskills (www.netskills.ac.uk) and the Internet Detective Tutorial (www.sosig.ac.uk/desire/internet-detective.html). The internet is not just about popstars and pornography. Some of the best resources for biologists are given in Table A1.

## Statistics software

While it is of vital importance to understand the basis for statistical tests, how and when to apply them and how to interpret the results, actually

**Table A2**   The three most popular statistical packages

| Package (supplier) | Advantages | Disadvantages |
|---|---|---|
| MINITAB (www.minitab.com) | Handles large datasets, reasonable range of statistical procedures | Complexity, confusing user interface. Read the manual, or *Practical Statistics for experimental Biologists*, by A.C. Wardlaw (Wiley, 2000) |
| SPSS | Handles very large datasets, very comprehensive range of statistical procedures | Complex user interface, although newer version are better. Probably no easier to use than MINITAB |
| Excel | Probably the easiest of the three to use, and you may have this software already. Can be programmed to do any statistical calculation if you know how | Primarily a spreadsheet, limited range of statistical procedures, online documentation poor |

performing all but the simplest calculations by hand is ridiculous. In a decade we have moved from reliance on pocket calculators to statistics software running on personal computers and the internet. There is a vast amount of statistics software available and, as with most software, what you use will depend largely on your local circumstances. A lot of useful statistics software is now freely available via the internet. For example, try searching on Google (www.google.com) for 'chi AND calculator', 'ANOVA AND calculator' or whatever you are interested in. In spite of this, there are still limitations of free online software and a free-standing statistical package will probably be needed. Far too many statistics packages are available to name them all, but three of the most popular are MINITAB®, SPSS® and Microsoft Excel® (Table A2).

# Appendix 3: Statistical Formulae and Tables

*Arithmetic mean* (Chapter 8)

$$\text{mean} = \frac{\sum X}{N}$$

where $\sum$ means 'sum', $X$ are the raw datapoints and $N$ is the number of scores.

*Binomial distribution* (Chapter 9):

$$(P + Q)^n$$

where $P$ is the probability of one of the possible events, $Q$ is the probability of the second event ($Q = 1 - P$) and $n$ is the number of trials in the series.

$\chi^2$ (Chapter 10):

$$\sum \frac{(\text{observed frequency} - \text{expected frequency})^2}{\text{expected frequency}}$$

*Deviation score* (Chapter 8):

$$X - m$$

where $X =$ raw score and $m =$ the mean for the dataset.

*F-ratio* (Chapter 10):

variance between groups/variance within sample groups.

*Interquartile range* (Chapter 8):

$$Q_3 - Q_1$$

*Linear regression* (Chapter 11):

$$y = a + bx$$

where

$$a = \frac{\sum y - b \sum x}{n}$$

$$b = \frac{n \sum (xy) - (\sum x)(\sum y)}{n \sum x^2 - (\sum x)^2}$$

*Paired t-test* (Chapter 10):

$$t = \frac{d_{av}}{SD/\sqrt{N}}$$

where $d_{av}$ is the mean difference, i.e. the sum of the differences of all the datapoints (set 1 point 1 − set 2 point 2, etc.) divided by the number of pairs, SD is the standard deviation of the differences between all the pairs and N is the number of pairs.

*Pearson correlation coefficient* (Chapter 11):

$$r = \frac{\sum XY - [(\sum X)(\sum Y)]/n}{\sqrt{\left\{\left[\sum X^2 - \frac{(\sum X)^2}{n}\right]\left[\sum Y^2 - \frac{(\sum Y)^2}{n}\right]\right\}}}$$

*Probability* (Chapter 9):

$$\frac{\text{number of specific outcomes of a trial}}{\text{total number of possible outcomes of a trial}}$$

*Range* (Chapter 8):

maximum value − minimum value

*Semi-interquartile range* (Chapter 8):

$$(Q_3 - Q_1)/2$$

*Spearman correlation coefficient* (Chapter 11):

$$1 - \frac{6 \sum D^2}{N(N^2 - 1)}$$

*Standard deviation* (Chapter 8):

Standard deviation of a population: $\sigma_x = \sqrt{\dfrac{\sum(X - \mu_x)^2}{N}}$

Standard deviation of a sample: $S_x = \sqrt{\dfrac{\sum(X - m)^2}{n - 1}}$

*Standard error* (Chapter 8):

$$SE = \frac{SD}{\sqrt{N}}$$

*Unpaired t-test* (Chapter 10):

$$t = \frac{\bar{x}_A - \bar{x}_B}{\sqrt{[(SE_A)^2 + (SE_B)^2]}}$$

where $\bar{x}$ is the mean of groups A and B, respectively, and $SE = SD/\sqrt{N}$.

*Variance of a population* (Chapter 8):

$$\sigma^2 = \frac{\sum(X - \mu_x)^2}{N}$$

where $\sum = $ sum, $X = $ raw score, $\mu_x = $ mean of the population and $N = $ number of datapoints.

*Variance of a sample* (Chapter 8):

$$S^2 = \frac{\sum(X - m)^2}{n}$$

where $\sum$ = sum, $X$ = raw score, $m$ = mean of the sample and $n$ = number of datapoints in the sample.

*z score* (Chapter 10):

$$(\text{score} - \text{mean})/\text{standard deviation}$$

# Critical values of the $\chi^2$ distribution

If the appropriate degree of freedom is not on this table, use the value for the next lower degree of freedom.

| df | | | | | $\alpha$ | | | | | |
|----|-------|-------|-------|-------|--------|--------|--------|--------|--------|--------|
|    | 0.995 | 0.99  | 0.975 | 0.95  | 0.9    | 0.1    | 0.05   | 0.025  | 0.01   | 0.005  |
| 1  | 0.000 | 0.000 | 0.001 | 0.004 | 0.016  | 2.706  | 3.841  | 5.024  | 6.635  | 7.879  |
| 2  | 0.010 | 0.020 | 0.051 | 0.103 | 0.211  | 4.605  | 5.991  | 7.378  | 9.210  | 10.597 |
| 3  | 0.072 | 0.115 | 0.216 | 0.352 | 0.584  | 6.251  | 7.815  | 9.348  | 11.345 | 12.838 |
| 4  | 0.207 | 0.297 | 0.484 | 0.711 | 1.064  | 7.779  | 9.488  | 11.143 | 13.277 | 14.860 |
| 5  | 0.412 | 0.554 | 0.831 | 1.145 | 1.610  | 9.236  | 11.070 | 12.832 | 15.086 | 16.750 |
| 6  | 0.676 | 0.872 | 1.237 | 1.635 | 2.204  | 10.645 | 12.592 | 14.449 | 16.812 | 18.548 |
| 7  | 0.989 | 1.239 | 1.690 | 2.167 | 2.833  | 12.017 | 14.067 | 16.013 | 18.475 | 20.278 |
| 8  | 1.344 | 1.647 | 2.180 | 2.733 | 3.490  | 13.362 | 15.507 | 17.535 | 20.090 | 21.955 |
| 9  | 1.735 | 2.088 | 2.700 | 3.325 | 4.168  | 14.684 | 16.919 | 19.023 | 21.666 | 23.589 |
| 10 | 2.156 | 2.558 | 3.247 | 3.940 | 4.865  | 15.987 | 18.307 | 20.483 | 23.209 | 25.188 |
| 11 | 2.603 | 3.053 | 3.816 | 4.575 | 5.578  | 17.275 | 19.675 | 21.920 | 24.725 | 26.757 |
| 12 | 3.074 | 3.571 | 4.404 | 5.226 | 6.304  | 18.549 | 21.026 | 23.337 | 26.217 | 28.300 |
| 13 | 3.565 | 4.107 | 5.009 | 5.892 | 7.041  | 19.812 | 22.362 | 24.736 | 27.688 | 29.819 |
| 14 | 4.075 | 4.660 | 5.629 | 6.571 | 7.790  | 21.064 | 23.685 | 26.119 | 29.141 | 31.319 |
| 15 | 4.601 | 5.229 | 6.262 | 7.261 | 8.547  | 22.307 | 24.996 | 27.488 | 30.578 | 32.801 |
| 16 | 5.142 | 5.812 | 6.908 | 7.962 | 9.312  | 23.542 | 26.296 | 28.845 | 32.000 | 34.267 |
| 17 | 5.697 | 6.408 | 7.564 | 8.672 | 10.085 | 24.769 | 27.587 | 30.191 | 33.409 | 35.718 |
| 18 | 6.265 | 7.015 | 8.231 | 9.390 | 10.865 | 25.989 | 28.869 | 31.526 | 34.805 | 37.156 |
| 19 | 6.844 | 7.633 | 8.907 | 10.117 | 11.651 | 27.204 | 30.144 | 32.852 | 36.191 | 38.582 |
| 20 | 7.434 | 8.260 | 9.591 | 10.851 | 12.443 | 28.412 | 31.410 | 34.170 | 37.566 | 39.997 |
| 21 | 8.034 | 8.897 | 10.283 | 11.591 | 13.240 | 29.615 | 32.671 | 35.479 | 38.932 | 41.401 |
| 22 | 8.643 | 9.542 | 10.982 | 12.338 | 14.041 | 30.813 | 33.924 | 36.781 | 40.289 | 42.796 |
| 23 | 9.260 | 10.196 | 11.689 | 13.091 | 14.848 | 32.007 | 35.172 | 38.076 | 41.638 | 44.181 |
| 24 | 9.886 | 10.856 | 12.401 | 13.848 | 15.659 | 33.196 | 36.415 | 39.364 | 42.980 | 45.558 |
| 25 | 10.520 | 11.524 | 13.120 | 14.611 | 16.473 | 34.382 | 37.652 | 40.646 | 44.314 | 46.928 |
| 26 | 11.160 | 12.198 | 13.844 | 15.379 | 17.292 | 35.563 | 38.885 | 41.923 | 45.642 | 48.290 |
| 27 | 11.808 | 12.878 | 14.573 | 16.151 | 18.114 | 36.741 | 40.113 | 43.195 | 46.963 | 49.645 |
| 28 | 12.461 | 13.565 | 15.308 | 16.928 | 18.939 | 37.916 | 41.337 | 44.461 | 48.278 | 50.994 |
| 29 | 13.121 | 14.256 | 16.047 | 17.708 | 19.768 | 39.087 | 42.557 | 45.722 | 49.588 | 52.335 |
| 30 | 13.787 | 14.953 | 16.791 | 18.493 | 20.599 | 40.256 | 43.773 | 46.979 | 50.892 | 53.672 |
| 31 | 14.458 | 15.655 | 17.539 | 19.281 | 21.434 | 41.422 | 44.985 | 48.232 | 52.191 | 55.002 |
| 32 | 15.134 | 16.362 | 18.291 | 20.072 | 22.271 | 42.585 | 46.194 | 49.480 | 53.486 | 56.328 |
| 33 | 15.815 | 17.073 | 19.047 | 20.867 | 23.110 | 43.745 | 47.400 | 50.725 | 54.775 | 57.648 |
| 34 | 16.501 | 17.789 | 19.806 | 21.664 | 23.952 | 44.903 | 48.602 | 51.966 | 56.061 | 58.964 |

Critical values of the $\chi^2$ distribution   (*continued*)

| df | 0.995 | 0.99 | 0.975 | 0.95 | 0.9 | 0.1 | 0.05 | 0.025 | 0.01 | 0.005 |
|----|-------|------|-------|------|-----|-----|------|-------|------|-------|
| 35 | 17.192 | 18.509 | 20.569 | 22.465 | 24.797 | 46.059 | 49.802 | 53.203 | 57.342 | 60.275 |
| 36 | 17.887 | 19.233 | 21.336 | 23.269 | 25.643 | 47.212 | 50.998 | 54.437 | 58.619 | 61.581 |
| 37 | 18.586 | 19.960 | 22.106 | 24.075 | 26.492 | 48.363 | 52.192 | 55.668 | 59.893 | 62.883 |
| 38 | 19.289 | 20.691 | 22.878 | 24.884 | 27.343 | 49.513 | 53.384 | 56.895 | 61.162 | 64.181 |
| 39 | 19.996 | 21.426 | 23.654 | 25.695 | 28.196 | 50.660 | 54.572 | 58.120 | 62.428 | 65.475 |
| 40 | 20.707 | 22.164 | 24.433 | 26.509 | 29.051 | 51.805 | 55.758 | 59.342 | 63.691 | 66.766 |
| 50 | 27.991 | 29.707 | 32.357 | 34.764 | 37.689 | 63.167 | 67.505 | 71.420 | 76.154 | 79.490 |
| 60 | 35.534 | 37.485 | 40.482 | 43.188 | 46.459 | 74.397 | 79.082 | 83.298 | 88.379 | 91.952 |
| 70 | 43.275 | 45.442 | 48.758 | 51.739 | 55.329 | 85.527 | 90.531 | 95.023 | 100.425 | 104.215 |
| 80 | 51.172 | 53.540 | 57.153 | 60.391 | 64.278 | 96.578 | 101.879 | 106.629 | 112.329 | 116.321 |
| 90 | 59.196 | 61.754 | 65.647 | 69.126 | 73.291 | 107.565 | 113.145 | 118.136 | 124.116 | 128.299 |
| 100 | 67.328 | 70.065 | 74.222 | 77.929 | 82.358 | 118.498 | 124.342 | 129.561 | 135.807 | 140.170 |
| 150 | 109.142 | 112.668 | 117.985 | 122.692 | 128.275 | 172.581 | 179.581 | 185.800 | 193.207 | 198.360 |
| 200 | 152.241 | 156.432 | 162.728 | 168.279 | 174.835 | 226.021 | 233.994 | 241.058 | 249.445 | 255.264 |

# Critical values of student's *t*-test

If the appropriate degree of freedom is not on this table, use the value for the next lower degree of freedom.

| df | $\alpha$, one tail | | | | | |
|----|-------|-------|-------|-------|-------|-------|
|    | 0.250 | 0.100 | 0.050 | 0.025 | 0.010 | 0.005 |
|    | | | $\alpha$, two tails | | | |
|    | 0.500 | 0.200 | 0.100 | 0.050 | 0.020 | 0.010 |
| 1 | 1.000 | 3.078 | 6.314 | 12.706 | 31.821 | 63.657 |
| 2 | 0.816 | 1.886 | 2.920 | 4.303 | 6.965 | 9.925 |
| 3 | 0.765 | 1.638 | 2.353 | 3.182 | 4.541 | 5.841 |
| 4 | 0.741 | 1.533 | 2.132 | 2.776 | 3.747 | 4.604 |
| 5 | 0.727 | 1.476 | 2.015 | 2.571 | 3.365 | 4.032 |
| 6 | 0.718 | 1.440 | 1.943 | 2.447 | 3.143 | 3.707 |
| 7 | 0.711 | 1.415 | 1.895 | 2.365 | 2.998 | 3.499 |
| 8 | 0.706 | 1.397 | 1.860 | 2.306 | 2.896 | 3.355 |
| 9 | 0.703 | 1.383 | 1.833 | 2.262 | 2.821 | 3.250 |
| 10 | 0.700 | 1.372 | 1.812 | 2.228 | 2.764 | 3.169 |
| 11 | 0.697 | 1.363 | 1.796 | 2.201 | 2.718 | 3.106 |
| 12 | 0.695 | 1.356 | 1.782 | 2.179 | 2.681 | 3.055 |
| 13 | 0.694 | 1.350 | 1.771 | 2.160 | 2.650 | 3.012 |
| 14 | 0.692 | 1.345 | 1.761 | 2.145 | 2.624 | 2.977 |
| 15 | 0.691 | 1.341 | 1.753 | 2.131 | 2.602 | 2.947 |
| 16 | 0.690 | 1.337 | 1.746 | 2.120 | 2.583 | 2.921 |
| 17 | 0.689 | 1.333 | 1.740 | 2.110 | 2.567 | 2.898 |

(*continued overleaf*)

Critical values of student's *t*-test   (*continued*)

| df | α, one tail | | | | | |
|----|-------|-------|-------|-------|-------|-------|
|    | 0.250 | 0.100 | 0.050 | 0.025 | 0.010 | 0.005 |
|    |       |       | α, two tails | | | |
|    | 0.500 | 0.200 | 0.100 | 0.050 | 0.020 | 0.010 |
| 18  | 0.688 | 1.330 | 1.734 | 2.101 | 2.552 | 2.878 |
| 19  | 0.688 | 1.328 | 1.729 | 2.093 | 2.539 | 2.861 |
| 20  | 0.687 | 1.325 | 1.725 | 2.086 | 2.528 | 2.845 |
| 21  | 0.686 | 1.323 | 1.721 | 2.080 | 2.518 | 2.831 |
| 22  | 0.686 | 1.321 | 1.717 | 2.074 | 2.508 | 2.819 |
| 23  | 0.685 | 1.319 | 1.714 | 2.069 | 2.500 | 2.807 |
| 24  | 0.685 | 1.318 | 1.711 | 2.064 | 2.492 | 2.797 |
| 25  | 0.684 | 1.316 | 1.708 | 2.060 | 2.485 | 2.787 |
| 26  | 0.684 | 1.315 | 1.706 | 2.056 | 2.479 | 2.779 |
| 27  | 0.684 | 1.314 | 1.703 | 2.052 | 2.473 | 2.771 |
| 28  | 0.683 | 1.313 | 1.701 | 2.048 | 2.467 | 2.763 |
| 29  | 0.683 | 1.311 | 1.699 | 2.045 | 2.462 | 2.756 |
| 30  | 0.683 | 1.310 | 1.697 | 2.042 | 2.457 | 2.750 |
| 40  | 0.681 | 1.303 | 1.684 | 2.021 | 2.423 | 2.704 |
| 50  | 0.679 | 1.299 | 1.676 | 2.009 | 2.403 | 2.678 |
| 60  | 0.679 | 1.296 | 1.671 | 2.000 | 2.390 | 2.660 |
| 70  | 0.678 | 1.294 | 1.667 | 1.994 | 2.381 | 2.648 |
| 80  | 0.678 | 1.292 | 1.664 | 1.990 | 2.374 | 2.639 |
| 90  | 0.677 | 1.291 | 1.662 | 1.987 | 2.368 | 2.632 |
| 100 | 0.677 | 1.290 | 1.660 | 1.984 | 2.364 | 2.626 |
| 200 | 0.674 | 1.282 | 1.645 | 1.960 | 2.326 | 2.576 |

# Table of critical values of the *F*-statistic

Within each cell, the first entry is the critical value of $F$ for the 0.05 significance level, and the second entry is the critical value for the 0.01 significance level.

| df denominator | df numerator | | | | | | | | | |
|----------------|-------|-------|-------|-------|-------|-------|-------|-------|-------|-------|
|                | 1     | 2     | 3     | 4     | 5     | 6     | 7     | 8     | 9     | 10    |
| 2  | 18.51 | 19.00 | 19.16 | 19.25 | 19.30 | 19.33 | 19.35 | 19.37 | 19.38 | 19.40 |
|    | 98.50 | 99.00 | 99.16 | 99.25 | 99.30 | 99.33 | 99.36 | 99.38 | 99.39 | 99.40 |
| 3  | 10.13 | 9.55  | 9.28  | 9.12  | 9.01  | 8.94  | 8.89  | 8.85  | 8.81  | 8.79  |
|    | 34.12 | 30.82 | 29.46 | 28.71 | 28.24 | 27.91 | 27.67 | 27.49 | 27.34 | 27.23 |
| 4  | 7.71  | 6.94  | 6.59  | 6.39  | 6.26  | 6.16  | 6.09  | 6.04  | 6.00  | 5.96  |
|    | 21.20 | 18.00 | 16.69 | 15.98 | 15.52 | 15.21 | 14.98 | 14.80 | 14.66 | 14.55 |
| 5  | 6.61  | 5.79  | 5.41  | 5.19  | 5.05  | 4.95  | 4.88  | 4.82  | 4.77  | 4.74  |
|    | 16.26 | 13.27 | 12.06 | 11.39 | 10.97 | 10.67 | 10.46 | 10.29 | 10.16 | 10.05 |

Table of critical values of the *F*-statistic  (*continued*)

| df denominator | df numerator | | | | | | | | | |
|---|---|---|---|---|---|---|---|---|---|---|
| | 1 | 2 | 3 | 4 | 5 | 6 | 7 | 8 | 9 | 10 |
| 6 | 5.99 | 5.14 | 4.76 | 4.53 | 4.39 | 4.28 | 4.21 | 4.15 | 4.10 | 4.06 |
| | 13.75 | 10.92 | 9.78 | 9.15 | 8.75 | 8.47 | 8.26 | 8.10 | 7.98 | 7.87 |
| 7 | 5.59 | 4.74 | 4.35 | 4.12 | 3.97 | 3.87 | 3.79 | 3.73 | 3.68 | 3.64 |
| | 12.25 | 9.55 | 8.45 | 7.85 | 7.46 | 7.19 | 6.99 | 6.84 | 6.72 | 6.62 |
| 8 | 5.32 | 4.46 | 4.07 | 3.84 | 3.69 | 3.58 | 3.50 | 3.44 | 3.39 | 3.35 |
| | 11.26 | 8.65 | 7.59 | 7.01 | 6.63 | 6.37 | 6.18 | 6.03 | 5.91 | 5.81 |
| 9 | 5.12 | 4.26 | 3.86 | 3.63 | 3.48 | 3.37 | 3.29 | 3.23 | 3.18 | 3.14 |
| | 10.56 | 8.02 | 6.99 | 6.42 | 6.06 | 5.80 | 5.61 | 5.47 | 5.35 | 5.26 |
| 10 | 4.96 | 4.10 | 3.71 | 3.48 | 3.33 | 3.22 | 3.14 | 3.07 | 3.02 | 2.98 |
| | 10.04 | 7.56 | 6.55 | 5.99 | 5.64 | 5.39 | 5.20 | 5.06 | 4.94 | 4.85 |
| 11 | 4.84 | 3.98 | 3.59 | 3.36 | 3.20 | 3.09 | 3.01 | 2.95 | 2.90 | 2.85 |
| | 9.65 | 7.21 | 6.22 | 5.67 | 5.32 | 5.07 | 4.89 | 4.74 | 4.63 | 4.54 |
| 12 | 4.75 | 3.89 | 3.49 | 3.26 | 3.11 | 3 | 2.91 | 2.85 | 2.80 | 2.75 |
| | 9.33 | 6.93 | 5.95 | 5.41 | 5.06 | 4.82 | 4.64 | 4.50 | 4.39 | 4.30 |
| 13 | 4.67 | 3.81 | 3.41 | 3.18 | 3.03 | 2.92 | 2.83 | 2.77 | 2.71 | 2.67 |
| | 9.07 | 6.70 | 5.74 | 5.21 | 4.86 | 4.62 | 4.44 | 4.30 | 4.19 | 4.10 |
| 14 | 4.60 | 3.74 | 3.34 | 3.11 | 2.96 | 2.85 | 2.76 | 2.70 | 2.65 | 2.60 |
| | 8.86 | 6.51 | 5.56 | 5.04 | 4.69 | 4.46 | 4.28 | 4.14 | 4.03 | 3.94 |
| 15 | 4.54 | 3.68 | 3.29 | 3.06 | 2.90 | 2.79 | 2.71 | 2.64 | 2.59 | 2.54 |
| | 8.68 | 6.36 | 5.42 | 4.89 | 4.56 | 4.32 | 4.14 | 4.00 | 3.89 | 3.80 |
| 16 | 4.49 | 3.63 | 3.24 | 3.01 | 2.85 | 2.74 | 2.66 | 2.59 | 2.54 | 2.49 |
| | 8.53 | 6.23 | 5.29 | 4.77 | 4.44 | 4.20 | 4.03 | 3.89 | 3.78 | 3.69 |
| 17 | 4.45 | 3.59 | 3.20 | 2.96 | 2.81 | 2.70 | 2.61 | 2.55 | 2.49 | 2.45 |
| | 8.40 | 6.11 | 5.19 | 4.67 | 4.34 | 4.10 | 3.93 | 3.79 | 3.68 | 3.59 |
| 18 | 4.41 | 3.55 | 3.16 | 2.93 | 2.77 | 2.66 | 2.58 | 2.51 | 2.46 | 2.41 |
| | 8.29 | 6.01 | 5.09 | 4.58 | 4.25 | 4.01 | 3.84 | 3.71 | 3.60 | 3.51 |
| 19 | 4.38 | 3.52 | 3.13 | 2.90 | 2.74 | 2.63 | 2.54 | 2.48 | 2.42 | 2.38 |
| | 8.18 | 5.93 | 5.01 | 4.50 | 4.17 | 3.94 | 3.77 | 3.63 | 3.52 | 3.43 |
| 20 | 4.35 | 3.49 | 3.10 | 2.87 | 2.71 | 2.60 | 2.51 | 2.45 | 2.39 | 2.35 |
| | 8.10 | 5.85 | 4.94 | 4.43 | 4.10 | 3.87 | 3.70 | 3.56 | 3.46 | 3.37 |
| 21 | 4.32 | 3.47 | 3.07 | 2.84 | 2.68 | 2.57 | 2.49 | 2.42 | 2.37 | 2.32 |
| | 8.02 | 5.78 | 4.87 | 4.37 | 4.04 | 3.81 | 3.64 | 3.51 | 3.40 | 3.31 |
| 22 | 4.30 | 3.44 | 3.05 | 2.82 | 2.66 | 2.55 | 2.46 | 2.40 | 2.34 | 2.30 |
| | 7.95 | 5.72 | 4.82 | 4.31 | 3.99 | 3.76 | 3.59 | 3.45 | 3.35 | 3.26 |
| 23 | 4.28 | 3.42 | 3.03 | 2.80 | 2.64 | 2.53 | 2.44 | 2.37 | 2.32 | 2.27 |
| | 7.88 | 5.66 | 4.76 | 4.26 | 3.94 | 3.71 | 3.54 | 3.41 | 3.30 | 3.21 |
| 24 | 4.26 | 3.40 | 3.01 | 2.78 | 2.62 | 2.51 | 2.42 | 2.36 | 2.30 | 2.25 |
| | 7.82 | 5.61 | 4.72 | 4.22 | 3.90 | 3.67 | 3.50 | 3.36 | 3.26 | 3.17 |
| 25 | 4.24 | 3.39 | 2.99 | 2.76 | 2.60 | 2.49 | 2.40 | 2.34 | 2.28 | 2.24 |
| | 7.77 | 5.57 | 4.68 | 4.18 | 3.85 | 3.63 | 3.46 | 3.32 | 3.22 | 3.13 |
| 26 | 4.23 | 3.37 | 2.98 | 2.74 | 2.59 | 2.47 | 2.39 | 2.32 | 2.27 | 2.22 |
| | 7.72 | 5.53 | 4.64 | 4.14 | 3.82 | 3.59 | 3.42 | 3.29 | 3.18 | 3.09 |
| 27 | 4.21 | 3.35 | 2.96 | 2.73 | 2.57 | 2.46 | 2.37 | 2.31 | 2.25 | 2.20 |
| | 7.68 | 5.49 | 4.60 | 4.11 | 3.78 | 3.56 | 3.39 | 3.26 | 3.15 | 3.06 |

(*continued overleaf*)

Table of critical values of the F-statistic   (*continued*)

| df denominator | 1 | 2 | 3 | 4 | 5 | 6 | 7 | 8 | 9 | 10 |
|---|---|---|---|---|---|---|---|---|---|---|
| | | | | | df numerator | | | | | |
| 28 | 4.20 | 3.34 | 2.95 | 2.71 | 2.56 | 2.45 | 2.36 | 2.29 | 2.24 | 2.19 |
| | 7.64 | 5.45 | 4.57 | 4.07 | 3.75 | 3.53 | 3.36 | 3.23 | 3.12 | 3.03 |
| 29 | 4.18 | 3.33 | 2.93 | 2.70 | 2.55 | 2.43 | 2.35 | 2.28 | 2.22 | 2.18 |
| | 7.60 | 5.42 | 4.54 | 4.04 | 3.73 | 3.50 | 3.33 | 3.20 | 3.09 | 3.00 |
| 30 | 4.17 | 3.32 | 2.92 | 2.69 | 2.53 | 2.42 | 2.33 | 2.27 | 2.21 | 2.16 |
| | 7.56 | 5.39 | 4.51 | 4.02 | 3.70 | 3.47 | 3.30 | 3.17 | 3.07 | 2.98 |
| 31 | 4.16 | 3.30 | 2.91 | 2.68 | 2.52 | 2.41 | 2.32 | 2.25 | 2.20 | 2.15 |
| | 7.53 | 5.36 | 4.48 | 3.99 | 3.67 | 3.45 | 3.28 | 3.15 | 3.04 | 2.96 |
| 32 | 4.15 | 3.29 | 2.90 | 2.67 | 2.51 | 2.40 | 2.31 | 2.24 | 2.19 | 2.14 |
| | 7.50 | 5.34 | 4.46 | 3.97 | 3.65 | 3.43 | 3.26 | 3.13 | 3.02 | 2.93 |
| 33 | 4.14 | 3.28 | 2.89 | 2.66 | 2.50 | 2.39 | 2.30 | 2.23 | 2.18 | 2.13 |
| | 7.47 | 5.31 | 4.44 | 3.95 | 3.63 | 3.41 | 3.24 | 3.11 | 3.00 | 2.91 |
| 34 | 4.13 | 3.28 | 2.88 | 2.65 | 2.49 | 2.38 | 2.29 | 2.23 | 2.17 | 2.12 |
| | 7.44 | 5.29 | 4.42 | 3.93 | 3.61 | 3.39 | 3.22 | 3.09 | 2.98 | 2.89 |
| 35 | 4.12 | 3.27 | 2.87 | 2.64 | 2.49 | 2.37 | 2.29 | 2.22 | 2.16 | 2.11 |
| | 7.42 | 5.27 | 4.40 | 3.91 | 3.59 | 3.37 | 3.20 | 3.07 | 2.96 | 2.88 |
| 36 | 4.11 | 3.26 | 2.87 | 2.63 | 2.48 | 2.36 | 2.28 | 2.21 | 2.15 | 2.11 |
| | 7.40 | 5.25 | 4.38 | 3.89 | 3.57 | 3.35 | 3.18 | 3.05 | 2.95 | 2.86 |
| 37 | 4.11 | 3.25 | 2.86 | 2.63 | 2.47 | 2.36 | 2.27 | 2.20 | 2.14 | 2.10 |
| | 7.37 | 5.23 | 4.36 | 3.87 | 3.56 | 3.33 | 3.17 | 3.04 | 2.93 | 2.84 |
| 38 | 4.10 | 3.24 | 2.85 | 2.62 | 2.46 | 2.35 | 2.26 | 2.19 | 2.14 | 2.09 |
| | 7.35 | 5.21 | 4.34 | 3.86 | 3.54 | 3.32 | 3.15 | 3.02 | 2.92 | 2.83 |
| 39 | 4.09 | 3.24 | 2.85 | 2.61 | 2.46 | 2.34 | 2.26 | 2.19 | 2.13 | 2.08 |
| | 7.33 | 5.19 | 4.33 | 3.84 | 3.53 | 3.30 | 3.14 | 3.01 | 2.90 | 2.81 |
| 40 | 4.08 | 3.23 | 2.84 | 2.61 | 2.45 | 2.34 | 2.25 | 2.18 | 2.12 | 2.08 |
| | 7.31 | 5.18 | 4.31 | 3.83 | 3.51 | 3.29 | 3.12 | 2.99 | 2.89 | 2.80 |
| 41 | 4.08 | 3.23 | 2.83 | 2.60 | 2.44 | 2.33 | 2.24 | 2.17 | 2.12 | 2.07 |
| | 7.30 | 5.16 | 4.30 | 3.81 | 3.50 | 3.28 | 3.11 | 2.98 | 2.87 | 2.79 |
| 42 | 4.07 | 3.22 | 2.83 | 2.59 | 2.44 | 2.32 | 2.24 | 2.17 | 2.11 | 2.06 |
| | 7.28 | 5.15 | 4.29 | 3.80 | 3.49 | 3.27 | 3.10 | 2.97 | 2.86 | 2.78 |
| 43 | 4.07 | 3.21 | 2.82 | 2.59 | 2.43 | 2.32 | 2.23 | 2.16 | 2.11 | 2.06 |
| | 7.26 | 5.14 | 4.27 | 3.79 | 3.48 | 3.25 | 3.09 | 2.96 | 2.85 | 2.76 |
| 44 | 4.06 | 3.21 | 2.82 | 2.58 | 2.43 | 2.31 | 2.23 | 2.16 | 2.10 | 2.05 |
| | 7.25 | 5.12 | 4.26 | 3.78 | 3.47 | 3.24 | 3.08 | 2.95 | 2.84 | 2.75 |
| 45 | 4.06 | 3.20 | 2.81 | 2.58 | 2.42 | 2.31 | 2.22 | 2.15 | 2.10 | 2.05 |
| | 7.23 | 5.11 | 4.25 | 3.77 | 3.45 | 3.23 | 3.07 | 2.94 | 2.83 | 2.74 |
| 46 | 4.05 | 3.20 | 2.81 | 2.57 | 2.42 | 2.30 | 2.22 | 2.15 | 2.09 | 2.04 |
| | 7.22 | 5.10 | 4.24 | 3.76 | 3.44 | 3.22 | 3.06 | 2.93 | 2.82 | 2.73 |
| 47 | 4.05 | 3.20 | 2.80 | 2.57 | 2.41 | 2.30 | 2.21 | 2.14 | 2.09 | 2.04 |
| | 7.21 | 5.09 | 4.23 | 3.75 | 3.43 | 3.21 | 3.05 | 2.92 | 2.81 | 2.72 |
| 48 | 4.04 | 3.19 | 2.80 | 2.57 | 2.41 | 2.29 | 2.21 | 2.14 | 2.08 | 2.03 |
| | 7.19 | 5.08 | 4.22 | 3.74 | 3.43 | 3.20 | 3.04 | 2.91 | 2.80 | 2.71 |
| 49 | 4.04 | 3.19 | 2.79 | 2.56 | 2.40 | 2.29 | 2.20 | 2.13 | 2.08 | 2.03 |
| | 7.18 | 5.07 | 4.21 | 3.73 | 3.42 | 3.19 | 3.03 | 2.90 | 2.79 | 2.71 |

Table of critical values of the F-statistic  (*continued*)

| df denominator | df numerator | | | | | | | | | |
|---|---|---|---|---|---|---|---|---|---|---|
| | 1 | 2 | 3 | 4 | 5 | 6 | 7 | 8 | 9 | 10 |
| 50 | 4.03 | 3.18 | 2.79 | 2.56 | 2.40 | 2.29 | 2.20 | 2.13 | 2.07 | 2.03 |
| | 7.17 | 5.06 | 4.20 | 3.72 | 3.41 | 3.19 | 3.02 | 2.89 | 2.78 | 2.70 |
| 51 | 4.03 | 3.18 | 2.79 | 2.55 | 2.40 | 2.28 | 2.20 | 2.13 | 2.07 | 2.02 |
| | 7.16 | 5.05 | 4.19 | 3.71 | 3.40 | 3.18 | 3.01 | 2.88 | 2.78 | 2.69 |
| 52 | 4.03 | 3.18 | 2.78 | 2.55 | 2.39 | 2.28 | 2.19 | 2.12 | 2.07 | 2.02 |
| | 7.15 | 5.04 | 4.18 | 3.70 | 3.39 | 3.17 | 3.00 | 2.87 | 2.77 | 2.68 |
| 53 | 4.02 | 3.17 | 2.78 | 2.55 | 2.39 | 2.28 | 2.19 | 2.12 | 2.06 | 2.01 |
| | 7.14 | 5.03 | 4.17 | 3.70 | 3.38 | 3.16 | 3.00 | 2.87 | 2.76 | 2.68 |
| 54 | 4.02 | 3.17 | 2.78 | 2.54 | 2.39 | 2.27 | 2.18 | 2.12 | 2.06 | 2.01 |
| | 7.13 | 5.02 | 4.17 | 3.69 | 3.38 | 3.16 | 2.99 | 2.86 | 2.76 | 2.67 |
| 55 | 4.02 | 3.16 | 2.77 | 2.54 | 2.38 | 2.27 | 2.18 | 2.11 | 2.06 | 2.01 |
| | 7.12 | 5.01 | 4.16 | 3.68 | 3.37 | 3.15 | 2.98 | 2.85 | 2.75 | 2.66 |
| 56 | 4.01 | 3.16 | 2.77 | 2.54 | 2.38 | 2.27 | 2.18 | 2.11 | 2.05 | 2.00 |
| | 7.11 | 5.01 | 4.15 | 3.67 | 3.36 | 3.14 | 2.98 | 2.85 | 2.74 | 2.66 |
| 57 | 4.01 | 3.16 | 2.77 | 2.53 | 2.38 | 2.26 | 2.18 | 2.11 | 2.05 | 2.00 |
| | 7.10 | 5.00 | 4.15 | 3.67 | 3.36 | 3.14 | 2.97 | 2.84 | 2.74 | 2.65 |
| 58 | 4.01 | 3.16 | 2.76 | 2.53 | 2.37 | 2.26 | 2.17 | 2.10 | 2.05 | 2.00 |
| | 7.09 | 4.99 | 4.14 | 3.66 | 3.35 | 3.13 | 2.96 | 2.83 | 2.73 | 2.64 |
| 59 | 4.00 | 3.15 | 2.76 | 2.53 | 2.37 | 2.26 | 2.17 | 2.10 | 2.04 | 2.00 |
| | 7.08 | 4.98 | 4.13 | 3.65 | 3.34 | 3.12 | 2.96 | 2.83 | 2.72 | 2.64 |
| 60 | 4.00 | 3.15 | 2.76 | 2.53 | 2.37 | 2.25 | 2.17 | 2.10 | 2.04 | 1.99 |
| | 7.08 | 4.98 | 4.13 | 3.65 | 3.34 | 3.12 | 2.95 | 2.82 | 2.72 | 2.63 |
| 61 | 4.00 | 3.15 | 2.76 | 2.52 | 2.37 | 2.25 | 2.16 | 2.09 | 2.04 | 1.99 |
| | 7.07 | 4.97 | 4.12 | 3.64 | 3.33 | 3.11 | 2.95 | 2.82 | 2.71 | 2.63 |
| 62 | 4.00 | 3.15 | 2.75 | 2.52 | 2.36 | 2.25 | 2.16 | 2.09 | 2.03 | 1.99 |
| | 7.06 | 4.96 | 4.11 | 3.64 | 3.33 | 3.11 | 2.94 | 2.81 | 2.71 | 2.62 |
| 63 | 3.99 | 3.14 | 2.75 | 2.52 | 2.36 | 2.25 | 2.16 | 2.09 | 2.03 | 1.98 |
| | 7.06 | 4.96 | 4.11 | 3.63 | 3.32 | 3.10 | 2.94 | 2.81 | 2.70 | 2.62 |
| 64 | 3.99 | 3.14 | 2.75 | 2.52 | 2.36 | 2.24 | 2.16 | 2.09 | 2.03 | 1.98 |
| | 7.05 | 4.95 | 4.10 | 3.63 | 3.32 | 3.10 | 2.93 | 2.80 | 2.70 | 2.61 |
| 65 | 3.99 | 3.14 | 2.75 | 2.51 | 2.36 | 2.24 | 2.15 | 2.08 | 2.03 | 1.98 |
| | 7.04 | 4.95 | 4.10 | 3.62 | 3.31 | 3.09 | 2.93 | 2.80 | 2.69 | 2.61 |
| 66 | 3.99 | 3.14 | 2.74 | 2.51 | 2.35 | 2.24 | 2.15 | 2.08 | 2.03 | 1.98 |
| | 7.04 | 4.94 | 4.09 | 3.62 | 3.31 | 3.09 | 2.92 | 2.79 | 2.69 | 2.60 |
| 67 | 3.98 | 3.13 | 2.74 | 2.51 | 2.35 | 2.24 | 2.15 | 2.08 | 2.02 | 1.98 |
| | 7.03 | 4.94 | 4.09 | 3.61 | 3.30 | 3.08 | 2.92 | 2.79 | 2.68 | 2.60 |
| 68 | 3.98 | 3.13 | 2.74 | 2.51 | 2.35 | 2.24 | 2.15 | 2.08 | 2.02 | 1.97 |
| | 7.02 | 4.93 | 4.08 | 3.61 | 3.30 | 3.08 | 2.91 | 2.78 | 2.68 | 2.59 |
| 69 | 3.98 | 3.13 | 2.74 | 2.50 | 2.35 | 2.23 | 2.15 | 2.08 | 2.02 | 1.97 |
| | 7.02 | 4.93 | 4.08 | 3.60 | 3.29 | 3.08 | 2.91 | 2.78 | 2.68 | 2.59 |
| 70 | 3.98 | 3.13 | 2.74 | 2.50 | 2.35 | 2.23 | 2.14 | 2.07 | 2.02 | 1.97 |
| | 7.01 | 4.92 | 4.07 | 3.60 | 3.29 | 3.07 | 2.91 | 2.78 | 2.67 | 2.59 |
| 71 | 3.98 | 3.13 | 2.73 | 2.50 | 2.34 | 2.23 | 2.14 | 2.07 | 2.01 | 1.97 |
| | 7.01 | 4.92 | 4.07 | 3.60 | 3.29 | 3.07 | 2.90 | 2.77 | 2.67 | 2.58 |

(*continued overleaf*)

Table of critical values of the *F*-statistic   (*continued*)

| df denominator | 1 | 2 | 3 | 4 | 5 | 6 | 7 | 8 | 9 | 10 |
|---|---|---|---|---|---|---|---|---|---|---|
| 72 | 3.97 | 3.12 | 2.73 | 2.50 | 2.34 | 2.23 | 2.14 | 2.07 | 2.01 | 1.96 |
|    | 7.00 | 4.91 | 4.07 | 3.59 | 3.28 | 3.06 | 2.90 | 2.77 | 2.66 | 2.58 |
| 73 | 3.97 | 3.12 | 2.73 | 2.50 | 2.34 | 2.23 | 2.14 | 2.07 | 2.01 | 1.96 |
|    | 7.00 | 4.91 | 4.06 | 3.59 | 3.28 | 3.06 | 2.89 | 2.77 | 2.66 | 2.57 |
| 74 | 3.97 | 3.12 | 2.73 | 2.50 | 2.34 | 2.22 | 2.14 | 2.07 | 2.01 | 1.96 |
|    | 6.99 | 4.90 | 4.06 | 3.58 | 3.28 | 3.06 | 2.89 | 2.76 | 2.66 | 2.57 |
| 75 | 3.97 | 3.12 | 2.73 | 2.49 | 2.34 | 2.22 | 2.13 | 2.06 | 2.01 | 1.96 |
|    | 6.99 | 4.90 | 4.05 | 3.58 | 3.27 | 3.05 | 2.89 | 2.76 | 2.65 | 2.57 |
| 76 | 3.97 | 3.12 | 2.72 | 2.49 | 2.33 | 2.22 | 2.13 | 2.06 | 2.01 | 1.96 |
|    | 6.98 | 4.90 | 4.05 | 3.58 | 3.27 | 3.05 | 2.88 | 2.75 | 2.65 | 2.56 |
| 77 | 3.97 | 3.12 | 2.72 | 2.49 | 2.33 | 2.22 | 2.13 | 2.06 | 2.00 | 1.96 |
|    | 6.98 | 4.89 | 4.05 | 3.57 | 3.26 | 3.05 | 2.88 | 2.75 | 2.65 | 2.56 |
| 78 | 3.96 | 3.11 | 2.72 | 2.49 | 2.33 | 2.22 | 2.13 | 2.06 | 2.00 | 1.95 |
|    | 6.97 | 4.89 | 4.04 | 3.57 | 3.26 | 3.04 | 2.88 | 2.75 | 2.64 | 2.56 |
| 79 | 3.96 | 3.11 | 2.72 | 2.49 | 2.33 | 2.22 | 2.13 | 2.06 | 2.00 | 1.95 |
|    | 6.97 | 4.88 | 4.04 | 3.57 | 3.26 | 3.04 | 2.87 | 2.75 | 2.64 | 2.55 |
| 80 | 3.96 | 3.11 | 2.72 | 2.49 | 2.33 | 2.21 | 2.13 | 2.06 | 2.00 | 1.95 |
|    | 6.96 | 4.88 | 4.04 | 3.56 | 3.26 | 3.04 | 2.87 | 2.74 | 2.64 | 2.55 |
| 81 | 3.96 | 3.11 | 2.72 | 2.48 | 2.33 | 2.21 | 2.12 | 2.05 | 2.00 | 1.95 |
|    | 6.96 | 4.88 | 4.03 | 3.56 | 3.25 | 3.03 | 2.87 | 2.74 | 2.63 | 2.55 |
| 82 | 3.96 | 3.11 | 2.72 | 2.48 | 2.33 | 2.21 | 2.12 | 2.05 | 2.00 | 1.95 |
|    | 6.95 | 4.87 | 4.03 | 3.56 | 3.25 | 3.03 | 2.87 | 2.74 | 2.63 | 2.54 |
| 83 | 3.96 | 3.11 | 2.71 | 2.48 | 2.32 | 2.21 | 2.12 | 2.05 | 1.99 | 1.95 |
|    | 6.95 | 4.87 | 4.03 | 3.55 | 3.25 | 3.03 | 2.86 | 2.73 | 2.63 | 2.54 |
| 84 | 3.95 | 3.11 | 2.71 | 2.48 | 2.32 | 2.21 | 2.12 | 2.05 | 1.99 | 1.95 |
|    | 6.95 | 4.87 | 4.02 | 3.55 | 3.24 | 3.02 | 2.86 | 2.73 | 2.63 | 2.54 |
| 85 | 3.95 | 3.10 | 2.71 | 2.48 | 2.32 | 2.21 | 2.12 | 2.05 | 1.99 | 1.94 |
|    | 6.94 | 4.86 | 4.02 | 3.55 | 3.24 | 3.02 | 2.86 | 2.73 | 2.62 | 2.54 |
| 86 | 3.95 | 3.10 | 2.71 | 2.48 | 2.32 | 2.21 | 2.12 | 2.05 | 1.99 | 1.94 |
|    | 6.94 | 4.86 | 4.02 | 3.55 | 3.24 | 3.02 | 2.85 | 2.73 | 2.62 | 2.53 |
| 87 | 3.95 | 3.10 | 2.71 | 2.48 | 2.32 | 2.20 | 2.12 | 2.05 | 1.99 | 1.94 |
|    | 6.94 | 4.86 | 4.02 | 3.54 | 3.24 | 3.02 | 2.85 | 2.72 | 2.62 | 2.53 |
| 88 | 3.95 | 3.10 | 2.71 | 2.48 | 2.32 | 2.20 | 2.12 | 2.05 | 1.99 | 1.94 |
|    | 6.93 | 4.85 | 4.01 | 3.54 | 3.23 | 3.01 | 2.85 | 2.72 | 2.62 | 2.53 |
| 89 | 3.95 | 3.10 | 2.71 | 2.47 | 2.32 | 2.20 | 2.11 | 2.04 | 1.99 | 1.94 |
|    | 6.93 | 4.85 | 4.01 | 3.54 | 3.23 | 3.01 | 2.85 | 2.72 | 2.61 | 2.53 |
| 90 | 3.95 | 3.10 | 2.71 | 2.47 | 2.32 | 2.20 | 2.11 | 2.04 | 1.99 | 1.94 |
|    | 6.93 | 4.85 | 4.01 | 3.53 | 3.23 | 3.01 | 2.84 | 2.72 | 2.61 | 2.52 |
| 91 | 3.95 | 3.10 | 2.70 | 2.47 | 2.31 | 2.20 | 2.11 | 2.04 | 1.98 | 1.94 |
|    | 6.92 | 4.85 | 4.00 | 3.53 | 3.23 | 3.01 | 2.84 | 2.71 | 2.61 | 2.52 |
| 92 | 3.94 | 3.10 | 2.70 | 2.47 | 2.31 | 2.20 | 2.11 | 2.04 | 1.98 | 1.94 |
|    | 6.92 | 4.84 | 4.00 | 3.53 | 3.22 | 3.00 | 2.84 | 2.71 | 2.61 | 2.52 |
| 93 | 3.94 | 3.09 | 2.70 | 2.47 | 2.31 | 2.20 | 2.11 | 2.04 | 1.98 | 1.93 |
|    | 6.92 | 4.84 | 4.00 | 3.53 | 3.22 | 3.00 | 2.84 | 2.71 | 2.60 | 2.52 |

Table of critical values of the F-statistic   (*continued*)

| df denominator | df numerator | | | | | | | | | |
|---|---|---|---|---|---|---|---|---|---|---|
| | 1 | 2 | 3 | 4 | 5 | 6 | 7 | 8 | 9 | 10 |
| 94 | 3.94 | 3.09 | 2.70 | 2.47 | 2.31 | 2.20 | 2.11 | 2.04 | 1.98 | 1.93 |
| | 6.91 | 4.84 | 4.00 | 3.53 | 3.22 | 3.00 | 2.84 | 2.71 | 2.60 | 2.52 |
| 95 | 3.94 | 3.09 | 2.70 | 2.47 | 2.31 | 2.20 | 2.11 | 2.04 | 1.98 | 1.93 |
| | 6.91 | 4.84 | 3.99 | 3.52 | 3.22 | 3.00 | 2.83 | 2.70 | 2.60 | 2.51 |
| 96 | 3.94 | 3.09 | 2.70 | 2.47 | 2.31 | 2.19 | 2.11 | 2.04 | 1.98 | 1.93 |
| | 6.91 | 4.83 | 3.99 | 3.52 | 3.21 | 3.00 | 2.83 | 2.70 | 2.60 | 2.51 |
| 97 | 3.94 | 3.09 | 2.70 | 2.47 | 2.31 | 2.19 | 2.11 | 2.04 | 1.98 | 1.93 |
| | 6.90 | 4.83 | 3.99 | 3.52 | 3.21 | 2.99 | 2.83 | 2.70 | 2.60 | 2.51 |
| 98 | 3.94 | 3.09 | 2.70 | 2.46 | 2.31 | 2.19 | 2.10 | 2.03 | 1.98 | 1.93 |
| | 6.90 | 4.83 | 3.99 | 3.52 | 3.21 | 2.99 | 2.83 | 2.70 | 2.59 | 2.51 |
| 99 | 3.94 | 3.09 | 2.70 | 2.46 | 2.31 | 2.19 | 2.10 | 2.03 | 1.98 | 1.93 |
| | 6.90 | 4.83 | 3.99 | 3.51 | 3.21 | 2.99 | 2.83 | 2.70 | 2.59 | 2.51 |
| 100 | 3.94 | 3.09 | 2.70 | 2.46 | 2.31 | 2.19 | 2.10 | 2.03 | 1.97 | 1.93 |
| | 6.90 | 4.82 | 3.98 | 3.51 | 3.21 | 2.99 | 2.82 | 2.69 | 2.59 | 2.50 |

# Table of critical values of the correlation coefficient, $r$

| df | Level of significance for a one-tailed test | | | |
|---|---|---|---|---|
| | 0.05 | 0.025 | 0.01 | 0.005 |
| | Level of significance for a two-tailed test | | | |
| | 0.10 | 0.05 | 0.02 | 0.01 |
| 1 | 0.988 | 0.997 | 0.9995 | 0.9999 |
| 2 | 0.900 | 0.950 | 0.980 | 0.990 |
| 3 | 0.805 | 0.878 | 0.934 | 0.959 |
| 4 | 0.729 | 0.811 | 0.882 | 0.917 |
| 5 | 0.669 | 0.754 | 0.833 | 0.874 |
| 6 | 0.622 | 0.707 | 0.789 | 0.834 |
| 7 | 0.582 | 0.666 | 0.750 | 0.798 |
| 8 | 0.549 | 0.632 | 0.716 | 0.765 |
| 9 | 0.521 | 0.602 | 0.685 | 0.735 |
| 10 | 0.497 | 0.576 | 0.658 | 0.708 |
| 11 | 0.476 | 0.553 | 0.634 | 0.684 |
| 12 | 0.458 | 0.532 | 0.612 | 0.661 |
| 13 | 0.441 | 0.514 | 0.592 | 0.641 |
| 14 | 0.426 | 0.497 | 0.574 | 0.628 |
| 15 | 0.412 | 0.482 | 0.558 | 0.606 |
| 16 | 0.400 | 0.468 | 0.542 | 0.590 |
| 17 | 0.389 | 0.456 | 0.528 | 0.575 |

(*continued overleaf*)

Table of critical values of the correlation coefficient, $r$   (*continued*)

| df | Level of significance for a one-tailed test | | | |
|---|---|---|---|---|
| | 0.05 | 0.025 | 0.01 | 0.005 |
| | | Level of significance for a two-tailed test | | |
| | 0.10 | 0.05 | 0.02 | 0.01 |
| 18 | 0.378 | 0.444 | 0.516 | 0.561 |
| 19 | 0.369 | 0.433 | 0.503 | 0.549 |
| 20 | 0.360 | 0.423 | 0.492 | 0.537 |
| 21 | 0.352 | 0.413 | 0.482 | 0.526 |
| 22 | 0.344 | 0.404 | 0.472 | 0.515 |
| 23 | 0.337 | 0.396 | 0.462 | 0.505 |
| 24 | 0.330 | 0.388 | 0.453 | 0.495 |
| 25 | 0.323 | 0.381 | 0.445 | 0.487 |
| 26 | 0.317 | 0.374 | 0.437 | 0.479 |
| 27 | 0.311 | 0.367 | 0.430 | 0.471 |
| 28 | 0.306 | 0.361 | 0.423 | 0.463 |
| 29 | 0.301 | 0.355 | 0.416 | 0.456 |
| 30 | 0.296 | 0.349 | 0.409 | 0.449 |
| 35 | 0.275 | 0.325 | 0.381 | 0.418 |
| 40 | 0.257 | 0.304 | 0.358 | 0.393 |
| 45 | 0.243 | 0.288 | 0.338 | 0.372 |
| 50 | 0.231 | 0.273 | 0.322 | 0.354 |
| 60 | 0.211 | 0.250 | 0.295 | 0.325 |
| 70 | 0.195 | 0.232 | 0.274 | 0.302 |
| 80 | 0.183 | 0.217 | 0.256 | 0.284 |
| 90 | 0.173 | 0.205 | 0.242 | 0.267 |
| 100 | 0.164 | 0.195 | 0.230 | 0.254 |

# Table of binomial probabilities

The following table lists the probability of attaining $s$ successes out of $n$ trials, where the probability of success in any one trial is $P$.

To find the probability of obtaining $s$ or more successes in $n$ trials, add the probabilities corresponding to $s, s+1, s+2, \ldots$ together. For example, the probability of obtaining one or more heads in two coin flips is $0.5 + 0.25 = 0.75$.

If $P$ is greater than 0.5, then use $1 - P$ as the probability and $n - s$ as the desired number of successes (i.e. look up the probability of obtaining $n - s$ *failures* rather than $s$ *successes*).

## Table of binomial probabilities

| n | s | 0.01 | 0.02 | 0.04 | 0.05 | 0.06 | 0.08 | 0.10 | 0.12 | 0.14 | 0.15 | 0.16 | 0.18 | 0.20 | 0.22 | 0.24 | 0.25 | 0.30 | 0.35 | 0.40 | 0.45 | 0.50 |
|---|---|------|------|------|------|------|------|------|------|------|------|------|------|------|------|------|------|------|------|------|------|------|
| 2 | 0 | 0.980 | 0.960 | 0.922 | 0.903 | 0.884 | 0.846 | 0.810 | 0.774 | 0.740 | 0.723 | 0.706 | 0.672 | 0.640 | 0.608 | 0.578 | 0.563 | 0.490 | 0.423 | 0.360 | 0.303 | 0.250 |
| 2 | 1 | 0.020 | 0.039 | 0.077 | 0.095 | 0.113 | 0.147 | 0.180 | 0.211 | 0.241 | 0.255 | 0.269 | 0.295 | 0.320 | 0.343 | 0.365 | 0.375 | 0.420 | 0.455 | 0.480 | 0.495 | 0.500 |
| 2 | 2 |  |  | 0.002 | 0.003 | 0.004 | 0.006 | 0.010 | 0.014 | 0.020 | 0.023 | 0.026 | 0.032 | 0.040 | 0.048 | 0.058 | 0.063 | 0.090 | 0.123 | 0.160 | 0.203 | 0.250 |
| 3 | 0 | 0.970 | 0.941 | 0.885 | 0.857 | 0.831 | 0.779 | 0.729 | 0.681 | 0.636 | 0.614 | 0.593 | 0.551 | 0.512 | 0.475 | 0.439 | 0.422 | 0.343 | 0.275 | 0.216 | 0.166 | 0.125 |
| 3 | 1 | 0.029 | 0.058 | 0.111 | 0.135 | 0.159 | 0.203 | 0.243 | 0.279 | 0.311 | 0.325 | 0.339 | 0.363 | 0.384 | 0.402 | 0.416 | 0.422 | 0.441 | 0.444 | 0.432 | 0.408 | 0.375 |
| 3 | 2 |  | 0.001 | 0.005 | 0.007 | 0.010 | 0.018 | 0.027 | 0.038 | 0.051 | 0.057 | 0.065 | 0.080 | 0.096 | 0.113 | 0.131 | 0.141 | 0.189 | 0.239 | 0.288 | 0.334 | 0.375 |
| 3 | 3 |  |  |  |  |  | 0.001 | 0.001 | 0.002 | 0.003 | 0.003 | 0.004 | 0.006 | 0.008 | 0.011 | 0.014 | 0.016 | 0.027 | 0.043 | 0.064 | 0.091 | 0.125 |
| 4 | 0 | 0.961 | 0.922 | 0.849 | 0.815 | 0.781 | 0.716 | 0.656 | 0.600 | 0.547 | 0.522 | 0.498 | 0.452 | 0.410 | 0.370 | 0.334 | 0.316 | 0.240 | 0.179 | 0.130 | 0.092 | 0.063 |
| 4 | 1 | 0.039 | 0.075 | 0.142 | 0.171 | 0.199 | 0.249 | 0.292 | 0.327 | 0.356 | 0.368 | 0.379 | 0.397 | 0.410 | 0.418 | 0.421 | 0.422 | 0.412 | 0.384 | 0.346 | 0.299 | 0.250 |
| 4 | 2 | 0.001 | 0.002 | 0.009 | 0.014 | 0.019 | 0.033 | 0.049 | 0.067 | 0.087 | 0.098 | 0.108 | 0.131 | 0.154 | 0.177 | 0.200 | 0.211 | 0.265 | 0.311 | 0.346 | 0.368 | 0.375 |
| 4 | 3 |  |  |  |  | 0.001 | 0.002 | 0.004 | 0.006 | 0.009 | 0.011 | 0.014 | 0.019 | 0.026 | 0.033 | 0.042 | 0.047 | 0.076 | 0.111 | 0.154 | 0.200 | 0.250 |
| 4 | 4 |  |  |  |  |  |  |  |  |  | 0.001 | 0.001 | 0.001 | 0.002 | 0.002 | 0.003 | 0.004 | 0.008 | 0.015 | 0.026 | 0.041 | 0.063 |
| 5 | 0 | 0.951 | 0.904 | 0.815 | 0.774 | 0.734 | 0.659 | 0.590 | 0.528 | 0.470 | 0.444 | 0.418 | 0.371 | 0.328 | 0.289 | 0.254 | 0.237 | 0.168 | 0.116 | 0.078 | 0.050 | 0.031 |
| 5 | 1 | 0.048 | 0.092 | 0.170 | 0.204 | 0.234 | 0.287 | 0.328 | 0.360 | 0.383 | 0.392 | 0.398 | 0.407 | 0.410 | 0.407 | 0.400 | 0.396 | 0.360 | 0.312 | 0.259 | 0.206 | 0.156 |
| 5 | 2 | 0.001 | 0.004 | 0.014 | 0.021 | 0.030 | 0.050 | 0.073 | 0.098 | 0.125 | 0.138 | 0.152 | 0.179 | 0.205 | 0.230 | 0.253 | 0.264 | 0.309 | 0.336 | 0.346 | 0.337 | 0.313 |
| 5 | 3 |  |  | 0.001 | 0.001 | 0.002 | 0.004 | 0.008 | 0.013 | 0.020 | 0.024 | 0.029 | 0.039 | 0.051 | 0.065 | 0.080 | 0.088 | 0.132 | 0.181 | 0.230 | 0.276 | 0.313 |
| 5 | 4 |  |  |  |  |  |  |  | 0.001 | 0.002 | 0.002 | 0.003 | 0.004 | 0.006 | 0.009 | 0.013 | 0.015 | 0.028 | 0.049 | 0.077 | 0.113 | 0.156 |
| 5 | 5 |  |  |  |  |  |  |  |  |  |  |  |  |  | 0.001 | 0.001 | 0.001 | 0.002 | 0.005 | 0.010 | 0.018 | 0.031 |
| 6 | 0 | 0.941 | 0.886 | 0.783 | 0.735 | 0.690 | 0.606 | 0.531 | 0.464 | 0.405 | 0.377 | 0.351 | 0.304 | 0.262 | 0.225 | 0.193 | 0.178 | 0.118 | 0.075 | 0.047 | 0.028 | 0.016 |
| 6 | 1 | 0.057 | 0.108 | 0.196 | 0.232 | 0.264 | 0.316 | 0.354 | 0.380 | 0.395 | 0.399 | 0.401 | 0.400 | 0.393 | 0.381 | 0.365 | 0.356 | 0.303 | 0.244 | 0.187 | 0.136 | 0.094 |
| 6 | 2 | 0.001 | 0.006 | 0.020 | 0.031 | 0.042 | 0.069 | 0.098 | 0.130 | 0.161 | 0.176 | 0.191 | 0.220 | 0.246 | 0.269 | 0.288 | 0.297 | 0.324 | 0.328 | 0.311 | 0.278 | 0.234 |
| 6 | 3 |  |  | 0.001 | 0.002 | 0.004 | 0.008 | 0.015 | 0.024 | 0.035 | 0.041 | 0.049 | 0.064 | 0.082 | 0.101 | 0.121 | 0.132 | 0.185 | 0.235 | 0.276 | 0.303 | 0.313 |
| 6 | 4 |  |  |  |  |  | 0.001 | 0.001 | 0.002 | 0.004 | 0.005 | 0.007 | 0.011 | 0.015 | 0.021 | 0.029 | 0.033 | 0.060 | 0.095 | 0.138 | 0.186 | 0.234 |
| 6 | 5 |  |  |  |  |  |  |  |  |  |  | 0.001 | 0.001 | 0.002 | 0.002 | 0.004 | 0.004 | 0.010 | 0.020 | 0.037 | 0.061 | 0.094 |
| 6 | 6 |  |  |  |  |  |  |  |  |  |  |  |  |  |  |  |  | 0.001 | 0.002 | 0.004 | 0.008 | 0.016 |
| 7 | 0 | 0.932 | 0.868 | 0.751 | 0.698 | 0.648 | 0.558 | 0.478 | 0.409 | 0.348 | 0.321 | 0.295 | 0.249 | 0.210 | 0.176 | 0.146 | 0.133 | 0.082 | 0.049 | 0.028 | 0.015 | 0.008 |
| 7 | 1 | 0.066 | 0.124 | 0.219 | 0.257 | 0.290 | 0.340 | 0.372 | 0.390 | 0.396 | 0.396 | 0.393 | 0.383 | 0.367 | 0.347 | 0.324 | 0.311 | 0.247 | 0.185 | 0.131 | 0.087 | 0.055 |

(continued overleaf)

Table of binomial probabilities  (*continued*)

| n | s | 0.01 | 0.02 | 0.04 | 0.05 | 0.06 | 0.08 | 0.10 | 0.12 | 0.14 | 0.15 | 0.16 | 0.18 | 0.20 | 0.22 | 0.24 | 0.25 | 0.30 | 0.35 | 0.40 | 0.45 | 0.50 |
|---|---|------|------|------|------|------|------|------|------|------|------|------|------|------|------|------|------|------|------|------|------|------|
| 7 | 2 | 0.002 | 0.008 | 0.027 | 0.041 | 0.055 | 0.089 | 0.124 | 0.160 | 0.194 | 0.210 | 0.225 | 0.252 | 0.275 | 0.293 | 0.307 | 0.311 | 0.318 | 0.298 | 0.261 | 0.214 | 0.164 |
| 7 | 3 |  |  | 0.002 | 0.004 | 0.006 | 0.013 | 0.023 | 0.036 | 0.053 | 0.062 | 0.071 | 0.092 | 0.115 | 0.138 | 0.161 | 0.173 | 0.227 | 0.268 | 0.290 | 0.292 | 0.273 |
| 7 | 4 |  |  |  |  |  | 0.001 | 0.003 | 0.005 | 0.009 | 0.011 | 0.014 | 0.020 | 0.029 | 0.039 | 0.051 | 0.058 | 0.097 | 0.144 | 0.194 | 0.239 | 0.273 |
| 7 | 5 |  |  |  |  |  |  |  |  | 0.001 | 0.001 | 0.002 | 0.003 | 0.004 | 0.007 | 0.010 | 0.012 | 0.025 | 0.047 | 0.077 | 0.117 | 0.164 |
| 7 | 6 |  |  |  |  |  |  |  |  |  |  |  |  |  | 0.001 | 0.001 | 0.001 | 0.004 | 0.008 | 0.017 | 0.032 | 0.055 |
| 7 | 7 |  |  |  |  |  |  |  |  |  |  |  |  |  |  |  |  |  | 0.001 | 0.002 | 0.004 | 0.008 |
| 8 | 0 | 0.923 | 0.851 | 0.721 | 0.663 | 0.610 | 0.513 | 0.430 | 0.360 | 0.299 | 0.272 | 0.248 | 0.204 | 0.168 | 0.137 | 0.111 | 0.100 | 0.058 | 0.032 | 0.017 | 0.008 | 0.004 |
| 8 | 1 | 0.075 | 0.139 | 0.240 | 0.279 | 0.311 | 0.357 | 0.383 | 0.392 | 0.390 | 0.385 | 0.378 | 0.359 | 0.336 | 0.309 | 0.281 | 0.267 | 0.198 | 0.137 | 0.090 | 0.055 | 0.031 |
| 8 | 2 | 0.003 | 0.010 | 0.035 | 0.051 | 0.070 | 0.109 | 0.149 | 0.187 | 0.222 | 0.238 | 0.252 | 0.276 | 0.294 | 0.305 | 0.311 | 0.311 | 0.296 | 0.259 | 0.209 | 0.157 | 0.109 |
| 8 | 3 |  |  | 0.003 | 0.005 | 0.009 | 0.019 | 0.033 | 0.051 | 0.072 | 0.084 | 0.096 | 0.121 | 0.147 | 0.172 | 0.196 | 0.208 | 0.254 | 0.279 | 0.279 | 0.257 | 0.219 |
| 8 | 4 |  |  |  |  | 0.001 | 0.002 | 0.005 | 0.009 | 0.015 | 0.018 | 0.023 | 0.033 | 0.046 | 0.061 | 0.077 | 0.087 | 0.136 | 0.188 | 0.232 | 0.263 | 0.273 |
| 8 | 5 |  |  |  |  |  |  |  | 0.001 | 0.002 | 0.003 | 0.003 | 0.006 | 0.009 | 0.014 | 0.020 | 0.023 | 0.047 | 0.081 | 0.124 | 0.172 | 0.219 |
| 8 | 6 |  |  |  |  |  |  |  |  |  |  |  | 0.001 | 0.001 | 0.002 | 0.003 | 0.004 | 0.010 | 0.022 | 0.041 | 0.070 | 0.109 |
| 8 | 7 |  |  |  |  |  |  |  |  |  |  |  |  |  |  |  |  | 0.001 | 0.003 | 0.008 | 0.016 | 0.031 |
| 8 | 8 |  |  |  |  |  |  |  |  |  |  |  |  |  |  |  |  |  |  | 0.001 | 0.002 | 0.004 |
| 9 | 0 | 0.914 | 0.834 | 0.693 | 0.630 | 0.573 | 0.472 | 0.387 | 0.316 | 0.257 | 0.232 | 0.208 | 0.168 | 0.134 | 0.107 | 0.085 | 0.075 | 0.040 | 0.021 | 0.010 | 0.005 | 0.002 |
| 9 | 1 | 0.083 | 0.153 | 0.260 | 0.299 | 0.329 | 0.370 | 0.387 | 0.388 | 0.377 | 0.368 | 0.357 | 0.331 | 0.302 | 0.271 | 0.240 | 0.225 | 0.156 | 0.100 | 0.060 | 0.034 | 0.018 |
| 9 | 2 | 0.003 | 0.013 | 0.043 | 0.063 | 0.084 | 0.129 | 0.172 | 0.212 | 0.245 | 0.260 | 0.272 | 0.291 | 0.302 | 0.306 | 0.304 | 0.300 | 0.267 | 0.216 | 0.161 | 0.111 | 0.070 |
| 9 | 3 |  | 0.001 | 0.004 | 0.008 | 0.013 | 0.026 | 0.045 | 0.067 | 0.093 | 0.107 | 0.121 | 0.149 | 0.176 | 0.201 | 0.224 | 0.234 | 0.267 | 0.272 | 0.251 | 0.212 | 0.164 |
| 9 | 4 |  |  |  | 0.001 | 0.001 | 0.003 | 0.007 | 0.014 | 0.023 | 0.028 | 0.035 | 0.049 | 0.066 | 0.085 | 0.106 | 0.117 | 0.172 | 0.219 | 0.251 | 0.260 | 0.246 |
| 9 | 5 |  |  |  |  |  |  | 0.001 | 0.002 | 0.004 | 0.005 | 0.007 | 0.011 | 0.017 | 0.024 | 0.033 | 0.039 | 0.074 | 0.118 | 0.167 | 0.213 | 0.246 |
| 9 | 6 |  |  |  |  |  |  |  |  |  | 0.001 | 0.001 | 0.002 | 0.003 | 0.005 | 0.007 | 0.009 | 0.021 | 0.042 | 0.074 | 0.116 | 0.164 |
| 9 | 7 |  |  |  |  |  |  |  |  |  |  |  |  |  | 0.001 | 0.001 | 0.001 | 0.004 | 0.010 | 0.021 | 0.041 | 0.070 |
| 9 | 8 |  |  |  |  |  |  |  |  |  |  |  |  |  |  |  |  |  | 0.001 | 0.004 | 0.008 | 0.018 |
| 9 | 9 |  |  |  |  |  |  |  |  |  |  |  |  |  |  |  |  |  |  |  | 0.001 | 0.002 |
| 10 | 0 | 0.904 | 0.817 | 0.665 | 0.599 | 0.539 | 0.434 | 0.349 | 0.279 | 0.221 | 0.197 | 0.175 | 0.137 | 0.107 | 0.083 | 0.064 | 0.056 | 0.028 | 0.013 | 0.006 | 0.003 | 0.001 |
| 10 | 1 | 0.091 | 0.167 | 0.277 | 0.315 | 0.344 | 0.378 | 0.387 | 0.380 | 0.360 | 0.347 | 0.333 | 0.302 | 0.268 | 0.235 | 0.203 | 0.188 | 0.121 | 0.072 | 0.040 | 0.021 | 0.010 |
| 10 | 2 | 0.004 | 0.015 | 0.052 | 0.075 | 0.099 | 0.148 | 0.194 | 0.233 | 0.264 | 0.276 | 0.286 | 0.298 | 0.302 | 0.298 | 0.288 | 0.282 | 0.233 | 0.176 | 0.121 | 0.076 | 0.044 |

# TABLE OF BINOMIAL PROBABILITIES

| n | r | 0.01 | 0.02 | 0.04 | 0.05 | 0.06 | 0.08 | 0.10 | 0.12 | 0.14 | 0.15 | 0.16 | 0.18 | 0.20 | 0.22 | 0.24 | 0.25 | 0.30 | 0.35 | 0.40 | 0.45 | 0.50 |
|---|---|---|---|---|---|---|---|---|---|---|---|---|---|---|---|---|---|---|---|---|---|---|
| 10 | 3 | | 0.001 | 0.006 | 0.010 | 0.017 | 0.034 | 0.057 | 0.085 | 0.115 | 0.130 | 0.145 | 0.174 | 0.201 | 0.224 | 0.243 | 0.250 | 0.267 | 0.252 | 0.215 | 0.166 | 0.117 |
| 10 | 4 | | | | 0.001 | 0.002 | 0.005 | 0.011 | 0.020 | 0.033 | 0.040 | 0.048 | 0.067 | 0.088 | 0.111 | 0.134 | 0.146 | 0.200 | 0.238 | 0.251 | 0.238 | 0.205 |
| 10 | 5 | | | | | | 0.001 | 0.001 | 0.003 | 0.006 | 0.008 | 0.011 | 0.018 | 0.026 | 0.037 | 0.051 | 0.058 | 0.103 | 0.154 | 0.201 | 0.234 | 0.246 |
| 10 | 6 | | | | | | | | | 0.001 | 0.001 | 0.002 | 0.003 | 0.005 | 0.009 | 0.013 | 0.016 | 0.037 | 0.069 | 0.111 | 0.160 | 0.205 |
| 10 | 7 | | | | | | | | | | | | | 0.001 | 0.001 | 0.002 | 0.003 | 0.009 | 0.021 | 0.042 | 0.075 | 0.117 |
| 10 | 8 | | | | | | | | | | | | | | | | | 0.001 | 0.004 | 0.011 | 0.023 | 0.044 |
| 10 | 9 | | | | | | | | | | | | | | | | | | 0.001 | 0.002 | 0.004 | 0.010 |
| 10 | 10 | | | | | | | | | | | | | | | | | | | | | 0.001 |
| 11 | 0 | 0.895 | 0.801 | 0.638 | 0.569 | 0.506 | 0.400 | 0.314 | 0.245 | 0.190 | 0.167 | 0.147 | 0.113 | 0.086 | 0.065 | 0.049 | 0.042 | 0.020 | 0.009 | 0.004 | 0.001 | |
| 11 | 1 | 0.099 | 0.180 | 0.293 | 0.329 | 0.355 | 0.382 | 0.384 | 0.368 | 0.341 | 0.325 | 0.308 | 0.272 | 0.236 | 0.202 | 0.170 | 0.155 | 0.093 | 0.052 | 0.027 | 0.013 | 0.005 |
| 11 | 2 | 0.005 | 0.018 | 0.061 | 0.087 | 0.113 | 0.166 | 0.213 | 0.251 | 0.277 | 0.287 | 0.293 | 0.299 | 0.295 | 0.284 | 0.268 | 0.258 | 0.200 | 0.140 | 0.089 | 0.051 | 0.027 |
| 11 | 3 | | 0.001 | 0.008 | 0.014 | 0.022 | 0.043 | 0.071 | 0.103 | 0.135 | 0.152 | 0.168 | 0.197 | 0.221 | 0.241 | 0.254 | 0.258 | 0.257 | 0.225 | 0.177 | 0.126 | 0.081 |
| 11 | 4 | | | 0.001 | 0.001 | 0.003 | 0.008 | 0.016 | 0.028 | 0.044 | 0.054 | 0.064 | 0.086 | 0.111 | 0.136 | 0.160 | 0.172 | 0.220 | 0.243 | 0.236 | 0.206 | 0.161 |
| 11 | 5 | | | | | | 0.001 | 0.002 | 0.005 | 0.010 | 0.013 | 0.017 | 0.027 | 0.039 | 0.054 | 0.071 | 0.080 | 0.132 | 0.183 | 0.221 | 0.236 | 0.226 |
| 11 | 6 | | | | | | | | 0.001 | 0.002 | 0.002 | 0.003 | 0.006 | 0.010 | 0.015 | 0.022 | 0.027 | 0.057 | 0.099 | 0.147 | 0.193 | 0.226 |
| 11 | 7 | | | | | | | | | | | | 0.001 | 0.002 | 0.003 | 0.005 | 0.006 | 0.017 | 0.038 | 0.070 | 0.113 | 0.161 |
| 11 | 8 | | | | | | | | | | | | | | | 0.001 | 0.001 | 0.004 | 0.010 | 0.023 | 0.046 | 0.081 |
| 11 | 9 | | | | | | | | | | | | | | | | | 0.001 | 0.002 | 0.005 | 0.013 | 0.027 |
| 11 | 10 | | | | | | | | | | | | | | | | | | | 0.001 | 0.002 | 0.005 |
| 12 | 0 | 0.886 | 0.785 | 0.613 | 0.540 | 0.476 | 0.368 | 0.282 | 0.216 | 0.164 | 0.142 | 0.123 | 0.092 | 0.069 | 0.051 | 0.037 | 0.032 | 0.014 | 0.006 | 0.002 | 0.001 | |
| 12 | 1 | 0.107 | 0.192 | 0.306 | 0.341 | 0.365 | 0.384 | 0.377 | 0.353 | 0.320 | 0.301 | 0.282 | 0.243 | 0.206 | 0.172 | 0.141 | 0.127 | 0.071 | 0.037 | 0.017 | 0.008 | 0.003 |
| 12 | 2 | 0.006 | 0.022 | 0.070 | 0.099 | 0.128 | 0.183 | 0.230 | 0.265 | 0.286 | 0.292 | 0.296 | 0.294 | 0.283 | 0.266 | 0.244 | 0.232 | 0.168 | 0.109 | 0.064 | 0.034 | 0.016 |
| 12 | 3 | | 0.001 | 0.010 | 0.017 | 0.027 | 0.053 | 0.085 | 0.120 | 0.155 | 0.172 | 0.188 | 0.215 | 0.236 | 0.250 | 0.257 | 0.258 | 0.240 | 0.195 | 0.142 | 0.092 | 0.054 |
| 12 | 4 | | | 0.001 | 0.002 | 0.004 | 0.010 | 0.021 | 0.037 | 0.057 | 0.068 | 0.080 | 0.106 | 0.133 | 0.159 | 0.183 | 0.194 | 0.231 | 0.237 | 0.213 | 0.170 | 0.121 |
| 12 | 5 | | | | | | 0.001 | 0.004 | 0.008 | 0.015 | 0.019 | 0.025 | 0.037 | 0.053 | 0.072 | 0.092 | 0.103 | 0.158 | 0.204 | 0.227 | 0.223 | 0.193 |
| 12 | 6 | | | | | | | | 0.001 | 0.003 | 0.004 | 0.005 | 0.010 | 0.016 | 0.024 | 0.034 | 0.040 | 0.079 | 0.128 | 0.177 | 0.212 | 0.226 |
| 12 | 7 | | | | | | | | | | 0.001 | 0.001 | 0.002 | 0.003 | 0.006 | 0.009 | 0.011 | 0.029 | 0.059 | 0.101 | 0.149 | 0.193 |
| 12 | 8 | | | | | | | | | | | | | 0.001 | 0.001 | 0.002 | 0.002 | 0.008 | 0.020 | 0.042 | 0.076 | 0.121 |
| 12 | 9 | | | | | | | | | | | | | | | | | 0.001 | 0.005 | 0.012 | 0.028 | 0.054 |
| 12 | 10 | | | | | | | | | | | | | | | | | | 0.001 | 0.002 | 0.007 | 0.016 |
| 12 | 11 | | | | | | | | | | | | | | | | | | | | 0.001 | 0.003 |
| 13 | 0 | 0.878 | 0.769 | 0.588 | 0.513 | 0.447 | 0.338 | 0.254 | 0.190 | 0.141 | 0.121 | 0.104 | 0.076 | 0.055 | 0.040 | 0.028 | 0.024 | 0.010 | 0.004 | 0.001 | | |
| 13 | 1 | 0.115 | 0.204 | 0.319 | 0.351 | 0.371 | 0.382 | 0.367 | 0.336 | 0.298 | 0.277 | 0.257 | 0.216 | 0.179 | 0.145 | 0.116 | 0.103 | 0.054 | 0.026 | 0.011 | 0.004 | 0.002 |

(continued overleaf)

Table of binomial probabilities   (continued)

P

| n | s | 0.01 | 0.02 | 0.04 | 0.05 | 0.06 | 0.08 | 0.10 | 0.12 | 0.14 | 0.15 | 0.16 | 0.18 | 0.20 | 0.22 | 0.24 | 0.25 | 0.30 | 0.35 | 0.40 | 0.45 | 0.50 |
|---|---|---|---|---|---|---|---|---|---|---|---|---|---|---|---|---|---|---|---|---|---|---|
| 13 | 2 | 0.007 | 0.025 | 0.080 | 0.111 | 0.142 | 0.199 | 0.245 | 0.275 | 0.291 | 0.294 | 0.293 | 0.285 | 0.268 | 0.245 | 0.220 | 0.206 | 0.139 | 0.084 | 0.045 | 0.022 | 0.010 |
| 13 | 3 |  | 0.002 | 0.012 | 0.021 | 0.033 | 0.064 | 0.100 | 0.138 | 0.174 | 0.190 | 0.205 | 0.229 | 0.246 | 0.254 | 0.254 | 0.252 | 0.218 | 0.165 | 0.111 | 0.066 | 0.035 |
| 13 | 4 |  |  | 0.001 | 0.003 | 0.005 | 0.014 | 0.028 | 0.047 | 0.071 | 0.084 | 0.098 | 0.126 | 0.154 | 0.179 | 0.201 | 0.210 | 0.234 | 0.222 | 0.184 | 0.135 | 0.087 |
| 13 | 5 |  |  |  |  | 0.001 | 0.002 | 0.006 | 0.012 | 0.021 | 0.027 | 0.033 | 0.050 | 0.069 | 0.091 | 0.114 | 0.126 | 0.180 | 0.215 | 0.221 | 0.199 | 0.157 |
| 13 | 6 |  |  |  |  |  |  | 0.001 | 0.002 | 0.004 | 0.006 | 0.008 | 0.015 | 0.023 | 0.034 | 0.048 | 0.056 | 0.103 | 0.155 | 0.197 | 0.217 | 0.209 |
| 13 | 7 |  |  |  |  |  |  |  |  | 0.001 | 0.001 | 0.002 | 0.003 | 0.006 | 0.010 | 0.015 | 0.019 | 0.044 | 0.083 | 0.131 | 0.177 | 0.209 |
| 13 | 8 |  |  |  |  |  |  |  |  |  |  |  | 0.001 | 0.001 | 0.002 | 0.004 | 0.005 | 0.014 | 0.034 | 0.066 | 0.109 | 0.157 |
| 13 | 9 |  |  |  |  |  |  |  |  |  |  |  |  |  |  | 0.001 | 0.001 | 0.003 | 0.010 | 0.024 | 0.050 | 0.087 |
| 13 | 10 |  |  |  |  |  |  |  |  |  |  |  |  |  |  |  |  | 0.001 | 0.002 | 0.006 | 0.016 | 0.035 |
| 13 | 11 |  |  |  |  |  |  |  |  |  |  |  |  |  |  |  |  |  |  | 0.001 | 0.004 | 0.010 |
| 13 | 12 |  |  |  |  |  |  |  |  |  |  |  |  |  |  |  |  |  |  |  |  | 0.002 |
| 14 | 0 | 0.869 | 0.754 | 0.565 | 0.488 | 0.421 | 0.311 | 0.229 | 0.167 | 0.121 | 0.103 | 0.087 | 0.062 | 0.044 | 0.031 | 0.021 | 0.018 | 0.007 | 0.002 | 0.001 |  |  |
| 14 | 1 | 0.123 | 0.215 | 0.329 | 0.359 | 0.376 | 0.379 | 0.356 | 0.319 | 0.276 | 0.254 | 0.232 | 0.191 | 0.154 | 0.122 | 0.095 | 0.083 | 0.041 | 0.018 | 0.007 | 0.003 | 0.001 |
| 14 | 2 | 0.008 | 0.029 | 0.089 | 0.123 | 0.156 | 0.214 | 0.257 | 0.283 | 0.292 | 0.291 | 0.287 | 0.272 | 0.250 | 0.223 | 0.195 | 0.180 | 0.113 | 0.063 | 0.032 | 0.014 | 0.006 |
| 14 | 3 |  | 0.002 | 0.015 | 0.026 | 0.040 | 0.074 | 0.114 | 0.154 | 0.190 | 0.206 | 0.219 | 0.239 | 0.250 | 0.252 | 0.246 | 0.240 | 0.194 | 0.137 | 0.085 | 0.046 | 0.022 |
| 14 | 4 |  |  | 0.002 | 0.004 | 0.007 | 0.018 | 0.035 | 0.058 | 0.085 | 0.100 | 0.115 | 0.144 | 0.172 | 0.195 | 0.214 | 0.220 | 0.229 | 0.202 | 0.155 | 0.104 | 0.061 |
| 14 | 5 |  |  |  |  | 0.001 | 0.003 | 0.008 | 0.016 | 0.028 | 0.035 | 0.044 | 0.063 | 0.086 | 0.110 | 0.135 | 0.147 | 0.196 | 0.218 | 0.207 | 0.170 | 0.122 |
| 14 | 6 |  |  |  |  |  |  | 0.001 | 0.003 | 0.007 | 0.009 | 0.012 | 0.021 | 0.032 | 0.047 | 0.064 | 0.073 | 0.126 | 0.176 | 0.207 | 0.209 | 0.183 |
| 14 | 7 |  |  |  |  |  |  |  |  | 0.001 | 0.002 | 0.003 | 0.005 | 0.009 | 0.015 | 0.023 | 0.028 | 0.062 | 0.108 | 0.157 | 0.195 | 0.209 |
| 14 | 8 |  |  |  |  |  |  |  |  |  |  |  | 0.001 | 0.002 | 0.004 | 0.006 | 0.008 | 0.023 | 0.051 | 0.092 | 0.140 | 0.183 |
| 14 | 9 |  |  |  |  |  |  |  |  |  |  |  |  |  | 0.001 | 0.001 | 0.002 | 0.007 | 0.018 | 0.041 | 0.076 | 0.122 |
| 14 | 10 |  |  |  |  |  |  |  |  |  |  |  |  |  |  |  |  | 0.001 | 0.005 | 0.014 | 0.031 | 0.061 |
| 14 | 11 |  |  |  |  |  |  |  |  |  |  |  |  |  |  |  |  |  | 0.001 | 0.003 | 0.009 | 0.022 |
| 14 | 12 |  |  |  |  |  |  |  |  |  |  |  |  |  |  |  |  |  |  | 0.001 | 0.002 | 0.006 |
| 14 | 13 |  |  |  |  |  |  |  |  |  |  |  |  |  |  |  |  |  |  |  |  | 0.001 |
| 15 | 0 | 0.860 | 0.739 | 0.542 | 0.463 | 0.395 | 0.286 | 0.206 | 0.147 | 0.104 | 0.087 | 0.073 | 0.051 | 0.035 | 0.024 | 0.016 | 0.013 | 0.005 | 0.002 |  |  |  |
| 15 | 1 | 0.130 | 0.226 | 0.339 | 0.366 | 0.378 | 0.373 | 0.343 | 0.301 | 0.254 | 0.231 | 0.209 | 0.168 | 0.132 | 0.102 | 0.077 | 0.067 | 0.031 | 0.013 | 0.005 | 0.002 |  |
| 15 | 2 | 0.009 | 0.032 | 0.099 | 0.135 | 0.169 | 0.227 | 0.267 | 0.287 | 0.290 | 0.286 | 0.279 | 0.258 | 0.231 | 0.201 | 0.171 | 0.156 | 0.092 | 0.048 | 0.022 | 0.009 | 0.003 |
| 15 | 3 |  | 0.003 | 0.018 | 0.031 | 0.047 | 0.086 | 0.129 | 0.170 | 0.204 | 0.218 | 0.230 | 0.245 | 0.250 | 0.246 | 0.234 | 0.225 | 0.170 | 0.111 | 0.063 | 0.032 | 0.014 |

# TABLE OF BINOMIAL PROBABILITIES

| n | x | | | | | | | | | | | | | | | | | | | |
|---|---|---|---|---|---|---|---|---|---|---|---|---|---|---|---|---|---|---|---|---|
| 15 | 4 | 0.002 | 0.005 | 0.009 | 0.022 | 0.043 | 0.069 | 0.100 | 0.116 | 0.131 | 0.162 | 0.188 | 0.208 | 0.221 | 0.225 | 0.219 | 0.179 | 0.127 | 0.078 | 0.042 |
| 15 | 5 | | 0.001 | 0.001 | 0.004 | 0.010 | 0.021 | 0.036 | 0.045 | 0.055 | 0.078 | 0.103 | 0.129 | 0.154 | 0.165 | 0.206 | 0.212 | 0.186 | 0.140 | 0.092 |
| 15 | 6 | | | | 0.001 | 0.002 | 0.005 | 0.010 | 0.013 | 0.017 | 0.029 | 0.043 | 0.061 | 0.081 | 0.092 | 0.147 | 0.191 | 0.207 | 0.191 | 0.153 |
| 15 | 7 | | | | | | 0.001 | 0.002 | 0.003 | 0.004 | 0.008 | 0.014 | 0.022 | 0.033 | 0.039 | 0.081 | 0.132 | 0.177 | 0.201 | 0.196 |
| 15 | 8 | | | | | | | | 0.001 | 0.001 | 0.002 | 0.003 | 0.006 | 0.010 | 0.013 | 0.035 | 0.071 | 0.118 | 0.165 | 0.196 |
| 15 | 9 | | | | | | | | | | | 0.001 | 0.001 | 0.003 | 0.003 | 0.012 | 0.030 | 0.061 | 0.105 | 0.153 |
| 15 | 10 | | | | | | | | | | | | | | 0.001 | 0.003 | 0.010 | 0.024 | 0.051 | 0.092 |
| 15 | 11 | | | | | | | | | | | | | | | 0.001 | 0.002 | 0.007 | 0.019 | 0.042 |
| 15 | 12 | | | | | | | | | | | | | | | | | 0.002 | 0.005 | 0.014 |
| 15 | 13 | | | | | | | | | | | | | | | | | | 0.001 | 0.003 |

# Appendix 4: Glossary

| | |
|---|---|
| accuracy | A measurement of how close the average of a set of measurements is to the true or target value. |
| algebra | The branch of mathematics in which symbols are used to represent numbers. |
| alternative hypothesis | The opposite of the null hypothesis. |
| ANOVA | Analysis of variance. |
| arithmetic | The mathematics of numbers. |
| Avogadro's number | The number of molecules in one mole. |
| build a fraction | The opposite of reducing a fraction – making a fraction more complex. |
| central tendency | A summary measure of the middle of a dataset. |
| chi-square test | (also chi-squared or $\chi^2$) A non-parametric test used to investigate whether the proportions of certain categories are different in different groups. |
| complex fraction | A fraction where the numerator, denominator or both contain a fraction. |
| compound fraction (or mixed number) | A fraction which contains integers and fractions. |
| confidence interval | An estimated range of values which is likely to include an unknown datapoint. |
| constants | Numbers or terms whose value is fixed. |
| continuous variables | The set of all values which consists of intervals, e.g. 0–9, 10–19, 20–29, etc. |

correlation | Demonstrates the degree to which two or more variables are related.

dalton | The mass of a molecule relative to one-twelfth of the mass of a carbon-12 atom.

dependent variables | Variables which depend on independent variables.

deviation score | A measure of by how much each point in a frequency distribution lies above or below the mean for the entire dataset.

dimension | An abstract quality of measurement without scale (e.g. length, mass, time, etc.).

discrete variables | The set of all possible values which consists only of isolated points, e.g. counting variables (1, 2, 3 …).

dispersion (or variability) | A measure of the amount of scatter in a dataset.

equation | A mathematical expression which contains an equal sign.

expression | A string of mathematical symbols which describes ('expresses') a (potential) calculation using operators and operands.

F ratio ('Fisher ratio') | Compares the variance within sample groups ('inherent variance') with the variance between groups ('treatment effect') and is the basis for ANOVA.

factor an integer | Break the integer down into a group of numbers whose product equals the original number.

force | An action which maintains or alters the position of a body, or distorts it.

formula | A mathematical term which represents a relationship between two or more variables and/or constants.

frequency distribution | The number of observations for each of the possible categories in a dataset.

independence | Where the result of the first sample does not affect the outcome of subsequent samples.

independent variables | Variables which are experimentally manipulated by an investigator in an experiment.

| | |
|---|---|
| integer | A whole number. |
| interquartile range | The interval between the 25th and 75th percentiles. |
| interval variables | Equally spaced variables without a true zero point. |
| kurtosis | The size of the 'tails' (extremities) of a frequency distribution. |
| mean | The average value of a dataset, i.e. the sum of all the data divided by the number of variables. |
| median | The middle value in a dataset, i.e. half the variables have values greater than the median and the other half values which are less. |
| mode | The most frequently occurring value in a dataset. |
| molecular weight | The mass of one mole of a substance (in grams per mole). |
| nominal variables | Variables with no inherent order or ranking sequence, e.g. numbers used as names. |
| non-parametric method | A statistical method which does not depend on the parameters of populations or probability distributions. |
| normal distribution | A symmetrical frequency distribution with most values concentrated in the centre of the range. |
| null hypothesis | A test hypothesis about a population parameter. |
| one-tailed test | Used where there is some basis (e.g. previous experimental observation) to predict the direction of the difference, for example, expectation of a significant difference between groups. |
| one-way ANOVA | An ANOVA test which tests the hypothesis that means from two or more samples are equal. |
| operands | Mathematical symbols which the operators act on. |
| operations | Mathematical process such as addition, subtraction, multiplication and division performed in a specified sequence and in accordance with specific rules. |

| | |
|---|---|
| operators | Mathematical symbols indicating an operation to be performed, e.g. plus, minus, divide, etc. |
| ordinal variables | Non-quantitative variables with an ordered series. |
| paired *t*-test | The version of the *t*-test used when each data point in one group corresponds to a matching data point in the other group. |
| parametric method | A statistical method which depends on the parameters of populations or probability distributions. |
| percentile | The point on a frequency distribution below which a specified percentage of cases in the distribution fall. |
| population | The entire group from which data may be collected and conclusions drawn. |
| power | The rate at which work is done, i.e. the amount of work per unit time. |
| precision | A measure of the closeness of repeated observations to each other without reference to the true or target value, i.e. the reproducibility of the result. |
| qualitative data | A set of observations where any single observation is a word or code that represents a class or category. |
| quantitative data | A set of observations where any single observation is a number that represents an amount or a count. |
| quartile | The 25th, 50th and 75th percentiles, which divide a dataset into quarters. |
| range | The interval from the highest datapoint to the lowest datapoint. |
| ratio variables | Variables spaced at equal intervals with a true zero point. |
| reduce | To simplify a fraction or expression to its simplest possible terms without changing its value. |
| regression | Demonstrates the relationship between selected values of $X$ and observed values of $Y$, from which the most probable value of $Y$ can be predicted for any value of $X$. |

| | |
|---|---|
| repeated measures ANOVA | An ANOVA test used when members of a random sample are measured under different conditions. |
| sample | A smaller group drawn from a population. |
| sampling fluctuation | How much a statistic varies from one sample to another. |
| scatter plot | A plot of the values of Y vs the corresponding values of X. |
| semi-interquartile range | Half of the interquartile range. |
| serial dilution | A series of small, accurate dilutions rather than a single large dilution. |
| simple fraction | A fraction where the numerator and the denominator are both integers. |
| simplify | Convert a mathematical expression to a simpler form containing fewer terms. |
| skew | Lack of symmetry of a frequency distribution about the mean. |
| solution | A homogeneous mixture where all the particles (the 'solute') exist as individual molecules or ions dissolved in a liquid (the 'solvent'). |
| solve | To find the values(s) of the variable(s) which make an equation true, i.e. both terms equal. |
| standard deviation | The square root of the variance. |
| standard error | The standard deviation of the sampling distribution of that statistic. |
| standard score | see $z$-score. |
| (Student's) $t$-test | A parametric statistical test used to compare two groups. |
| tail(s) | Region(s) of rejection on the distribution of a test statistic. |
| two-tailed test | Used where there is no basis to assume that there may be a significant difference between the groups. |
| two-way ANOVA | An ANOVA test which simultaneously tests the hypothesis that the means of two variables from two or more groups are equal. |
| unit | A number which specifies a previously agreed scale. |

| | |
|---|---|
| unpaired *t*-test | The version of the *t*-test which can be used whether or not the groups contain matching datapoints. |
| variability (or dispersion) | A measure of the amount of scatter in a dataset. |
| variable | A quantity with no fixed value, capable of assuming any of a set of values. |
| variance | The mean of all the deviation scores for a dataset, i.e. the amount of deviation of the entire dataset from the mean. |
| work | The transfer of energy from one system to another. |
| *z*-score | A statistic which defines the position of a score in relation to the mean using the standard deviation as a unit of measurement. |

# Index